The Laboratory Mouse

The Laboratory Mouse

A Guide to the Location and Orientation of Tissues for Optimal Histological Evaluation

By
Jennifer Johnson, BS, HTL(ASCP)
Brian DelGiudice, BS
Dinesh S. Bangari, BVSc & AH, MS, PhD, Diplomate ACVP
Eleanor Peterson, HT(ASCP)
Gregory Ulinski, MS
Susan Ryan, MS, HT(ASCP)HTL
Beth L. Thurberg, MD, PhD
Edited by Gayle Callis, HT(ASCP)HTL, MT(ASCP)

CRC Press is an imprint of the
Taylor & Francis Group, an **informa** business

CRC Press
Taylor & Francis Group
6000 Broken Sound Parkway NW, Suite 300
Boca Raton, FL 33487-2742

Printed on acid-free paper
Version Date: 20190215

International Standard Book Number-13: 978-0-367-17800-0 (Hardback)

International Standard Book Number-13: 978-0-367-17775-1 (Paperback)

Library of Congress Cataloging-in-Publication Data

Names: Johnson, Jennifer (Staff scientist), author. | Callis, Gayle, editor.
Title: The laboratory mouse : a guide to the location and orientation of tissues for optimal histological evaluation / by Jennifer Johnson, Brian DelGiudice, Dinesh S. Bangari, Eleanor Peterson, Gregory Ulinski, Susan Ryan, Beth L. Thurberg ; edited by Gayle Callis.
Description: Boca Raton, Florida : CRC Press, [2019] | Includes bibliographical references and index.
Identifiers: LCCN 2018053218| ISBN 9780367178000 (hardback : alk. paper) | ISBN 9780367177751 (pbk. : alk. paper) | ISBN 9780429057755 (e-book)
Subjects: | MESH: Mice--anatomy & histology | Animals, Laboratory--anatomy & histology | Tissues--anatomy & histology
Classification: LCC SF407.M5 | NLM QY 60.R6 | DDC 616.02/7333--dc23
LC record available at https://lccn.loc.gov/2018053218

Visit the Taylor & Francis Web site at
http://www.taylorandfrancis.com

and the CRC Press Web site at
http://www.crcpress.com

Table of Contents

Preface

The purpose of this book has evolved over the process of its completion. It was originally intended to be an internal document to be used as a guide for the trimming, embedding, and orientation of the most commonly submitted mouse tissues handled by the laboratory. These pages would show the histologist how to orient the tissues and what the section should look like for the required pathology analysis. The guide would be a set of departmental standards to reference if mouse tissues were submitted without specific instructions. In its original format, the book would have served as a reference to ensure consistent treatment of mouse tissues dispite different research objectives.

To meet these needs, I consulted with pathologists, read textbooks and poured over countless web sites. I consulted the histologists who were trimming, processing, embedding, cutting, and staining the tissues. As I did this research, I realized there was a piece missing. There was nothing I could find that showed the complete connection from the mouse to the microscope. I could not find any references that connected the location and orientation in the mouse with the trimming and orientation of the organs to the final product — the perfect microscope slide that contained the elements a pathologist wants in order to evaluate a specific tissue!

I decided that if I wanted to fill that gap, the focus of the book would need to be expanded! That is what you will find on the pages to follow.

I hope that this book will help those who receive their mouse tissues already in a jar of formalin and need guidance on embedding and creating sections. I want them to look at the left-hand page and see the images showing how organs reside within the mouse. This will help readers more fully appreciate why a particular orientation of a specific organ is so important. I organized this book to be a useful guide for those doing mouse necropsy to not only to show where to locate but also how to collect an organ. After collecting the organ, look at the page on the right for trimming, orientation during embedding, facing into a tissue, and seeing a good H&E section of the organ's components. I want technicians to understand how proper collection techniques will affect all the processes downstream. How an organ is collected affects every step from trimming through staining of the section to have a beautiful hematoxylin and eosin section for microscopic examination. This book is all about making the connections from one step to the next to gain a better understanding of murine tissue collection and produce the best possible results.

And this labor would not have been possible without the help, patience and dedication of all of the talented authors, contributors and the editor. It has been years in the making, and I hope that you, the reader, find it a valuable resource.

Jennifer Johnson BS, HTL(ASCP)

The authors would like to acknowledge Linda Agee-Suarez, Yingli Yang, Kristen Legendre, and Peter Piepenhagen for staining, sectioning, necropsy support and slide scanning. A special thank you to Gayle Callis for her time and expertise in editing this book.

Disclosure: At the time this book was written, all authors were employees of Sanofi, Framingham, MA 01701.

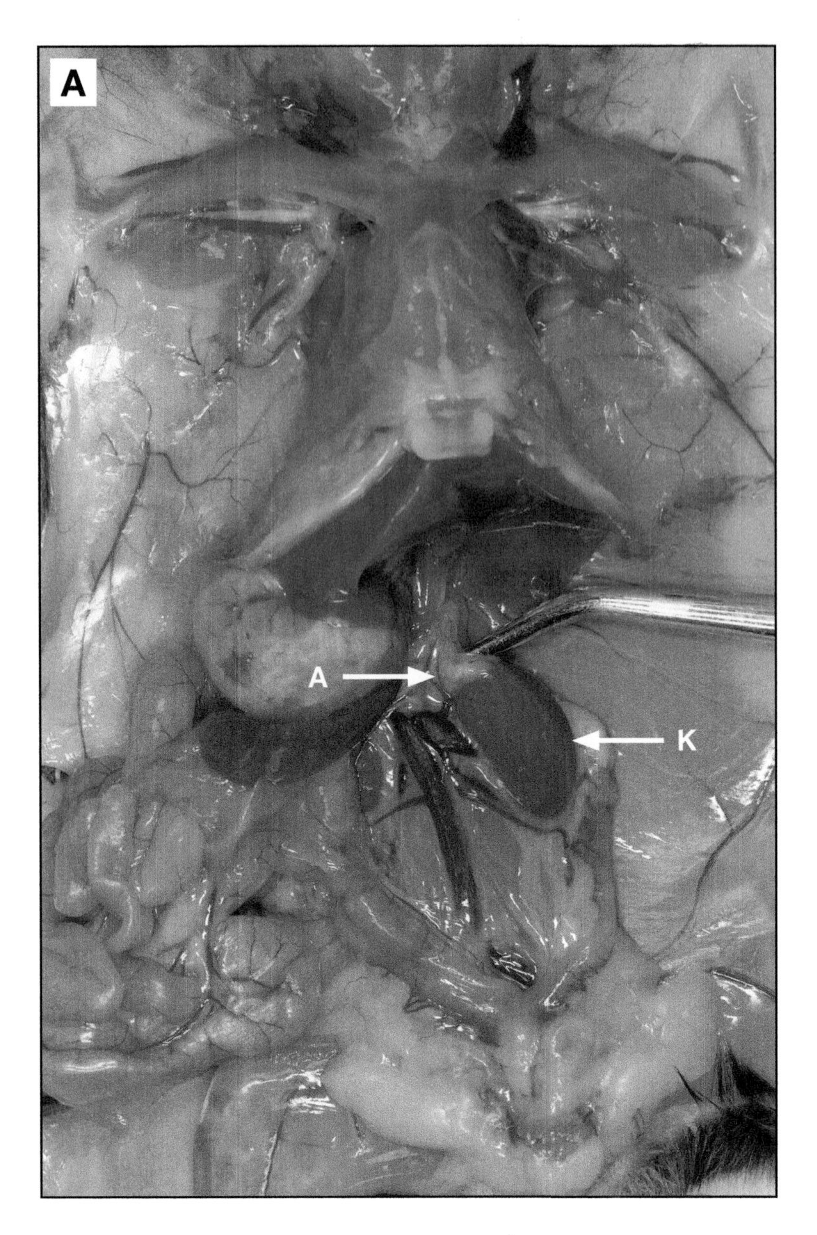

Adrenal Glands

A. One adrenal gland (A) is located at the cranial pole of each kidney (K). The gland's orange to pink color helps distinguish it from the surrounding fat. To remove the adrenal gland, grasp the surrounding fat, not the fragile gland to avoid forceps artifact. Fat must be carefully removed if the glands are to be weighed or used for biochemical analysis. Fat removal does not hinder pathological analysis.

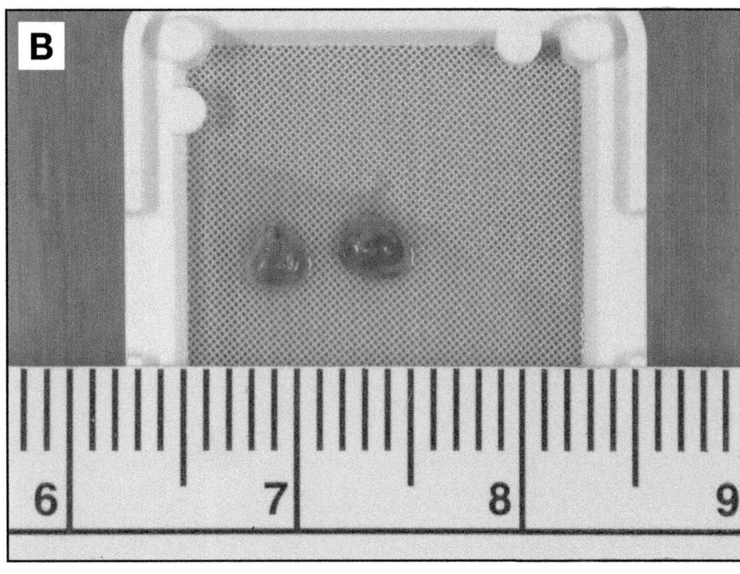

B. Secure the glands in a CellSafe™ Biopsy Insert or a piece of paper towel prior to cassetting them.

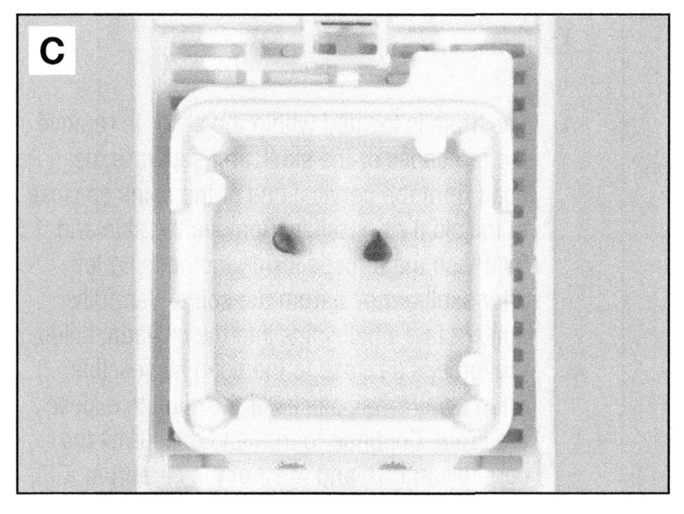

C. After processing, the adrenal glands will appear smaller and darker.

D. Embed the adrenal glands as close to the bottom of the embedding mold as possible.

E. When facing into the glands, try to get into both adrenal glands.

F. A section of adult mouse adrenal showing the cortex (C) and medulla (M) should be obtained.

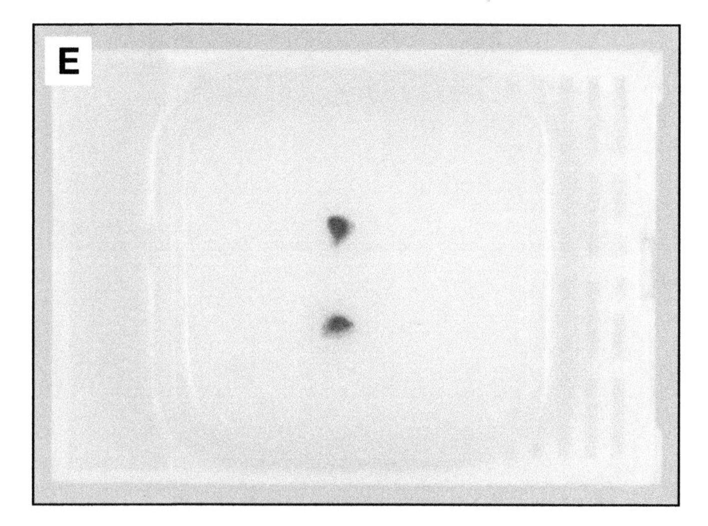

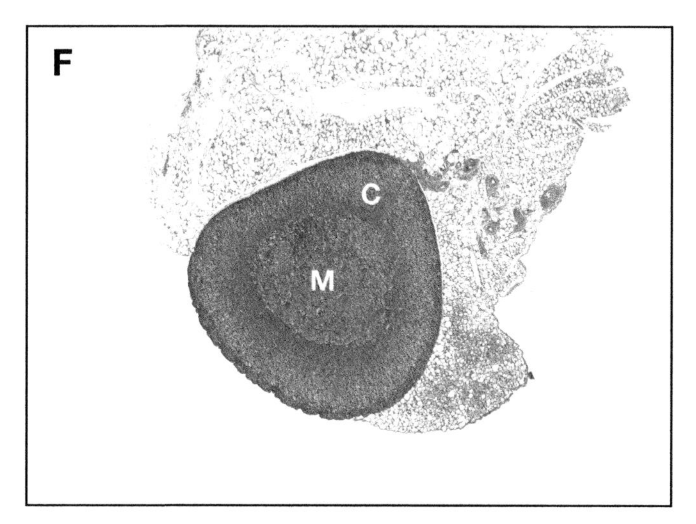

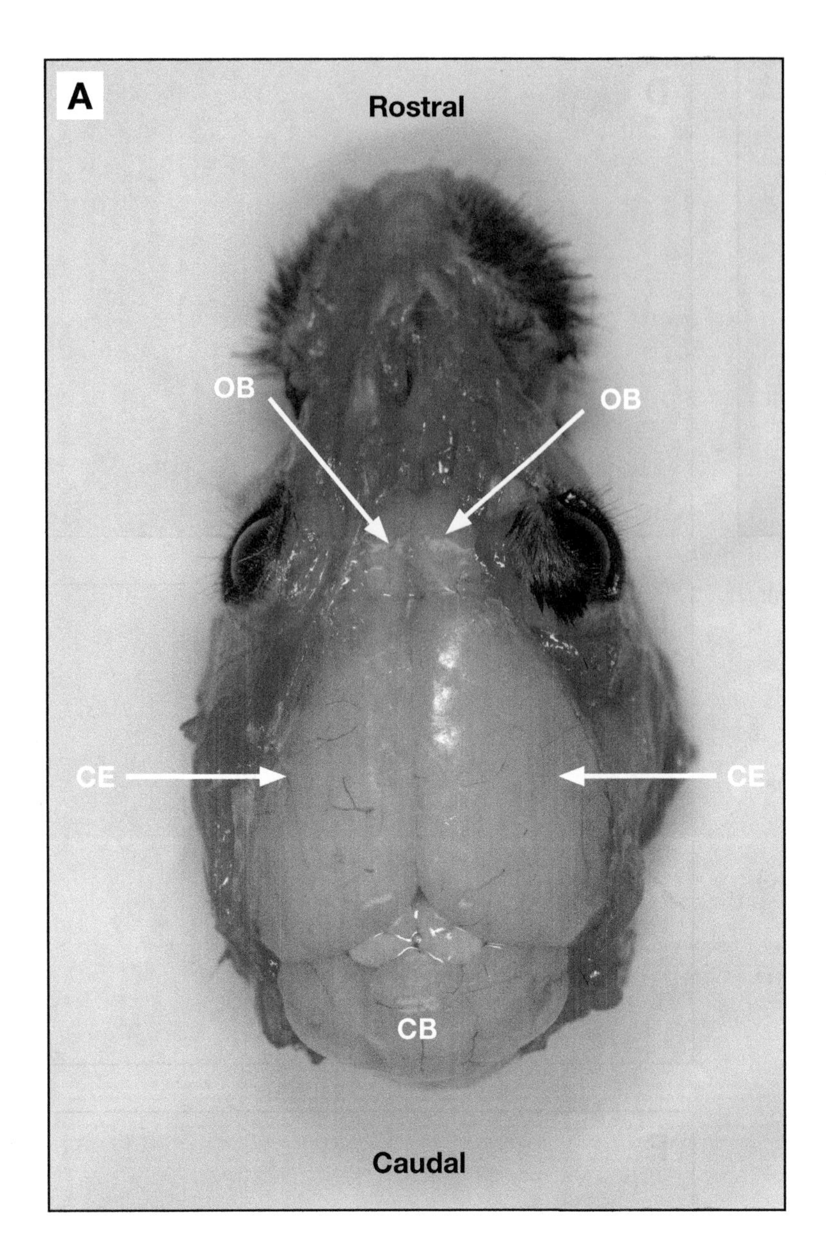

Brain

A. The brain is located within the skull. To remove the top bones of the skull, first separate the head from the cervical end of the spine as close to the skull as possible. Remove the skin and fur. Place the forceps inside the opening left after skull removal from the spine. Carefully make a few small snips into the calvaria, holding the forceps as close to the bone as possible. Snap away small pieces of the skull to expose the brain. Continue to trim and snap until the cerebellum (CB) and most of the cerebrum (CE) are exposed. For best results, fix the brain inside the open skull for 24 hours prior to removing it completely. After fixation, carefully dislodge the brain using a small metal weigh spoon. Take care to preserve the olfactory bulbs (OB).

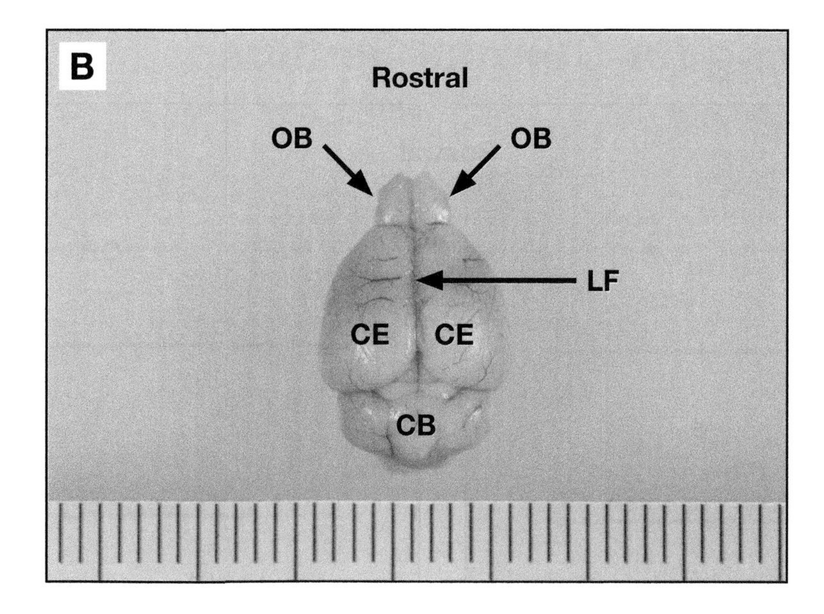

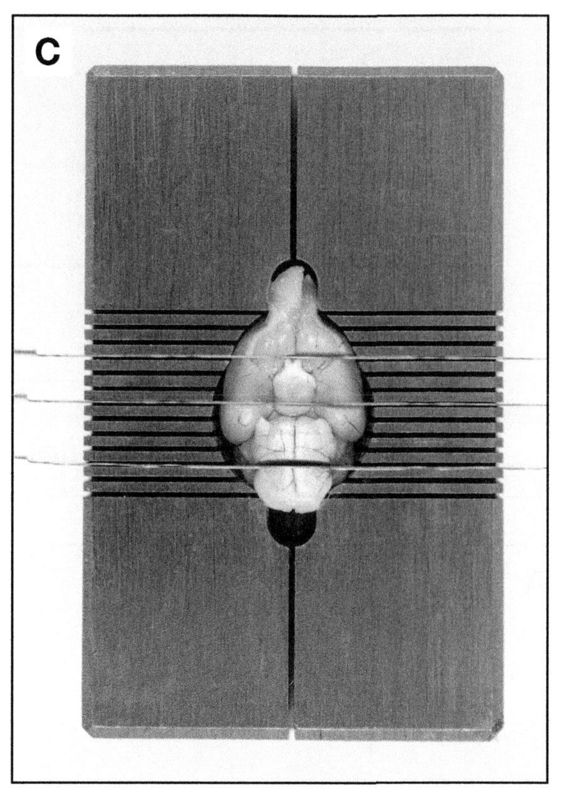

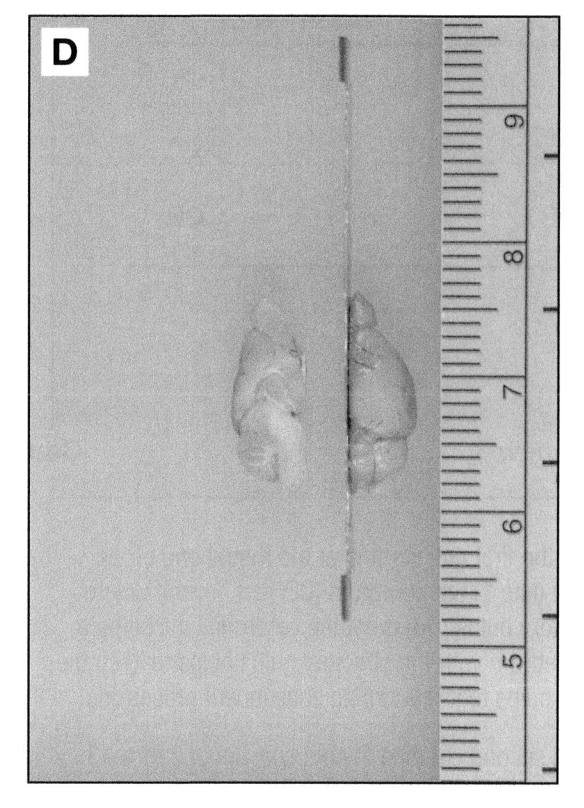

B. Dorsal View: After fixation, the brain is firmer and easier to handle during trimming. Here the olfactory bulbs (OB), cerebrum (CE) and cerebellum (CB) are well preserved. The longitudinal fissure (LF) separates the brain into the left and right hemispheres.

C. Ventral View: To obtain consistent coronal sections, a rodent brain matrix is used. Place the brain into the matrix ventral side up. Do not force the brain into the matrix to avoid tissue damage. Because every brain is a little different, it is important to use anatomic cues when placing the blades.

D. Dorsal View: Cutting the brain into sagittal sections can be done in a matrix, but is more easily done by cutting along the line of the longitudinal fissure.

Brain - Trimming for Coronal Sections

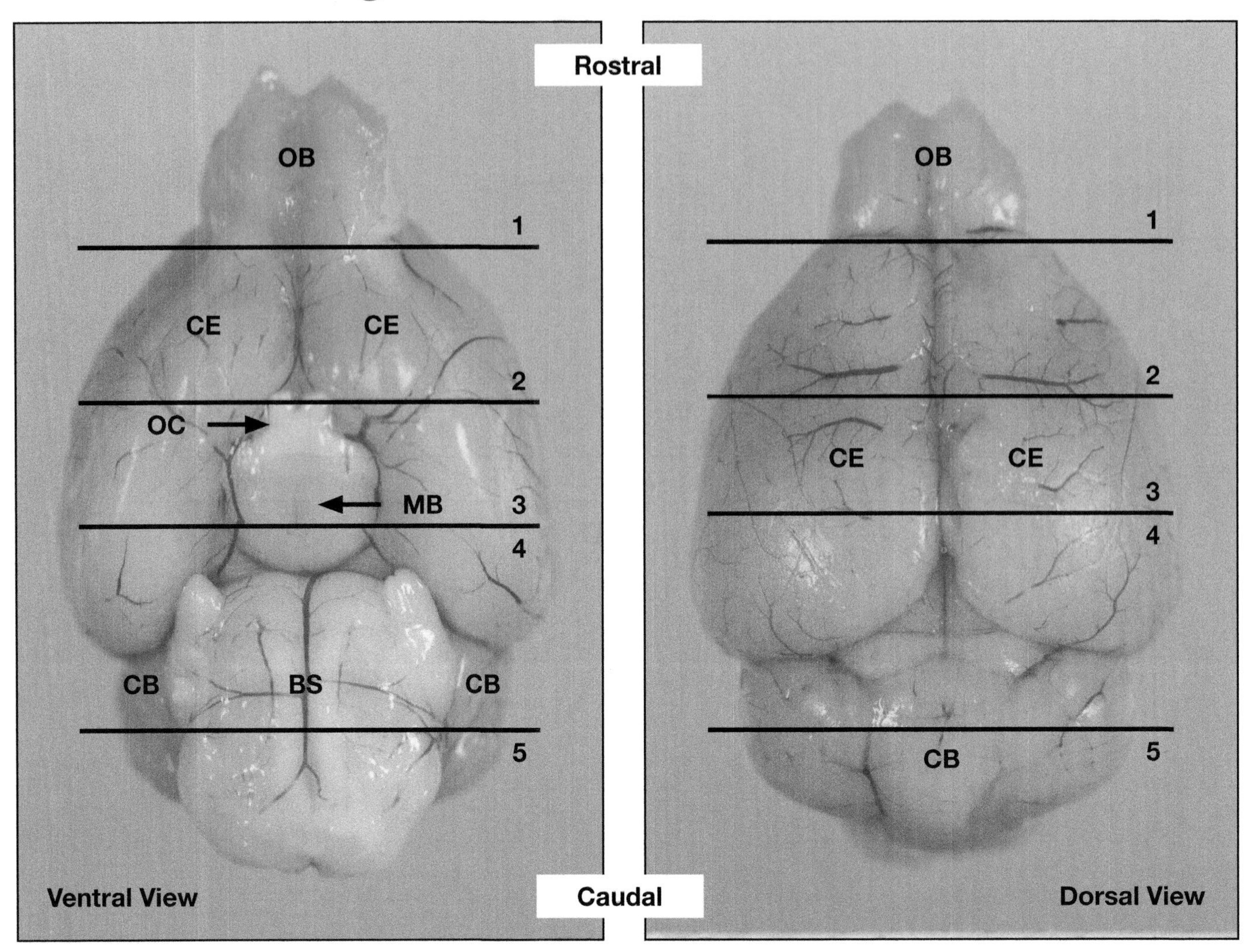

1. To make the first cut, starting at the rostral end of the brain, cut (line 1) the cerebrum (CE) just beyond where the olfactory bulbs (OB) meet the cerebrum. Including a bit of cerebrum with the olfactory bulbs helps to keep the olfactory bulbs together during subsequent processing.
2. Make the second cut (line 2) coronally about 2 mm - 3 mm into the brain at the level of the optic chiasma (OC). The caudal face will be placed downward into the cassette for sectioning.
3. To make the cut at line 3/4, place the blade at the posterior 1/3 of the mammillary body (MB). Place the newly cut face (caudal side) down into the cassette.
4. The final cut (line 5) is made just rostral to the widest part of the cerebellum. This section is placed rostral face down into the cassette.
5. The remaining piece of brain – cerebellum (CB) and brain stem (BS) is placed cut face down into the cassette for sectioning.

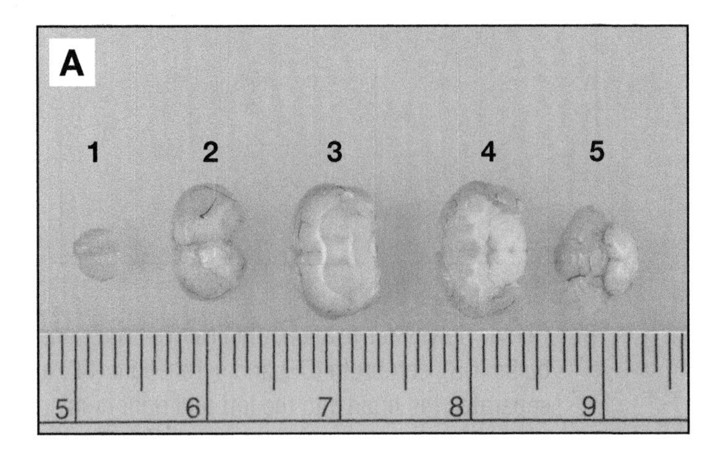

A. Five coronal brain sections after trimming for correct orientation and placement into the cassette. Cover with a thin sponge or a small piece of paper towel to hold them in place.

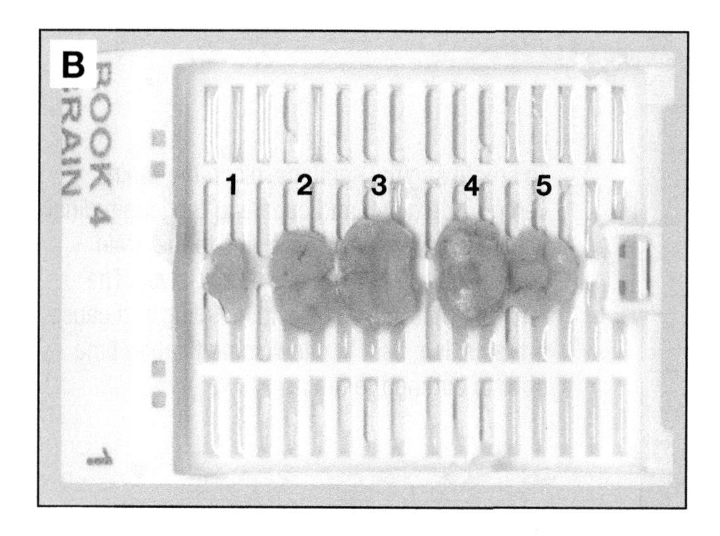

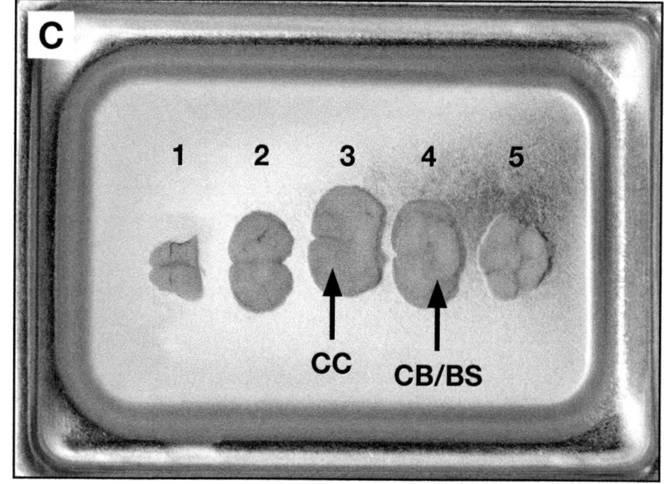

B. After processing, the coronal pieces will be smaller and more fragile. It is common for the olfactory bulbs to separate during processing.

C. Embed the coronal sections as they are sitting in the cassette. Be sure to press them gently but firmly into the bottom of the embedding mold. If the pieces have flipped during processing, use the photos above and the guide to embed them in order:

   Slice 1 – Place the long edges of the olfactory bulbs flat into the embedding mold so that both bulbs can be sectioned.

   Slice 2 – This piece contains the front of the cerebrum from which the olfactory bulbs have been removed. Place piece 2 with the rounded side up and the flat surface down into the mold.

   Slice 3 – This is the thickest of the remaining pieces. Place piece 3 with the white line of the corpus callosum (CC) facing up as shown.

   Slice 4 – This piece contains the back of the cerebrum and the beginning of the cerebellum and brain stem (CB/BS). The cerebellum and brain stem side should face up.

   Slice 5 – This piece has a rounded face and a flat face. The flat face is placed down into the embedding mold.

D. A technician facing the block, should see full faces of all five coronal slices.

E. The ideal sections should demonstrate the following areas of the brain: olfactory bulbs (OB), corpus callosum (CC), basal ganglia (BG), thalamus (T), hypothalamus (HT), hippocampus (HC), cerebellum (CB), and brain stem (BS).

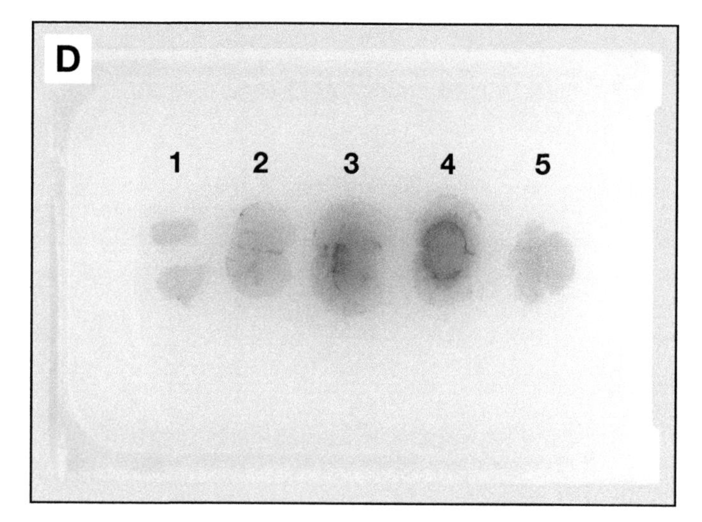

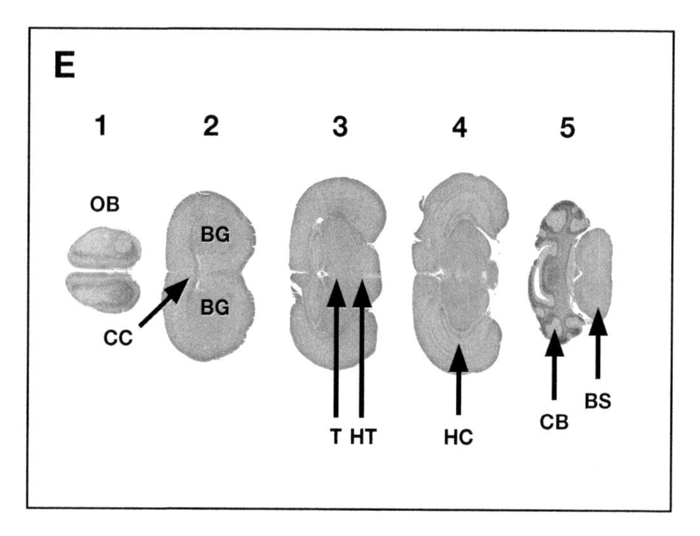

Brain - Trimming for Sagittal Sections

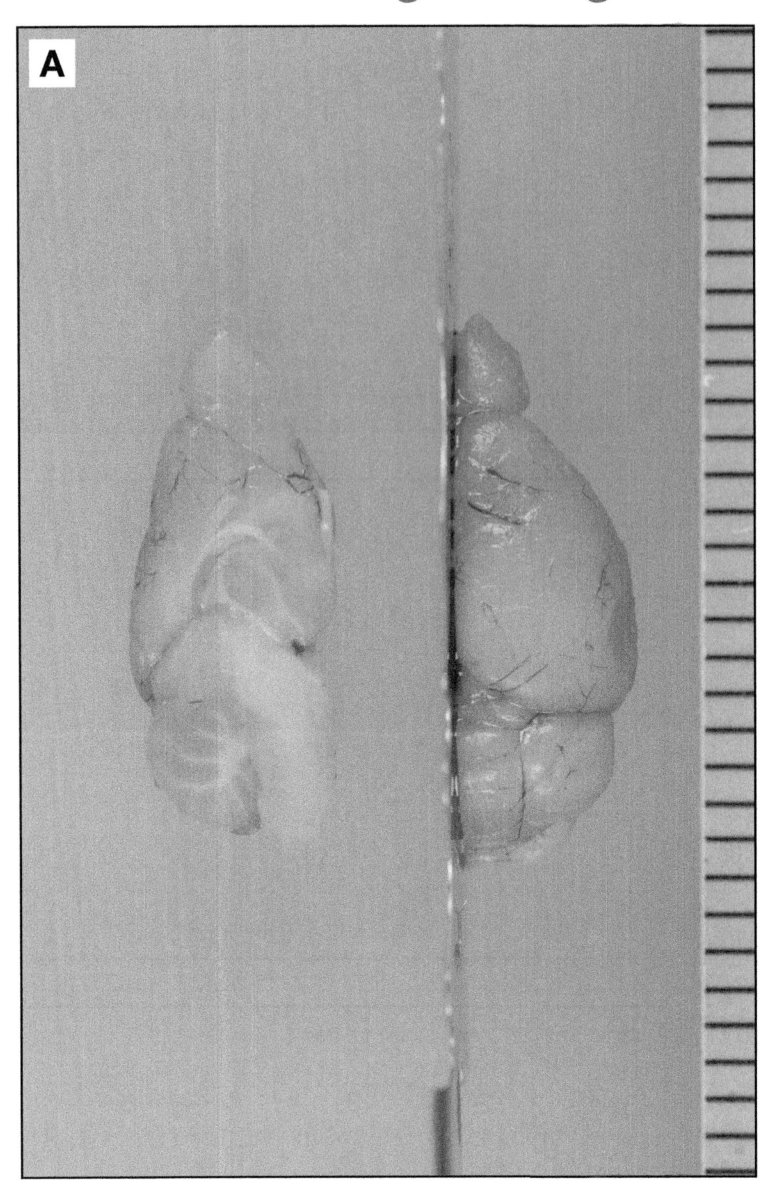

A. To create sagittal sections, place the brain ventral side down and cut along the longitudinal fissure from the olfactory bulbs to the brain stem to create two equally sized halves. The fatty brain tissue is soft and handling can cause artifacts. It is best to allow some fixation time prior to cutting the brain.

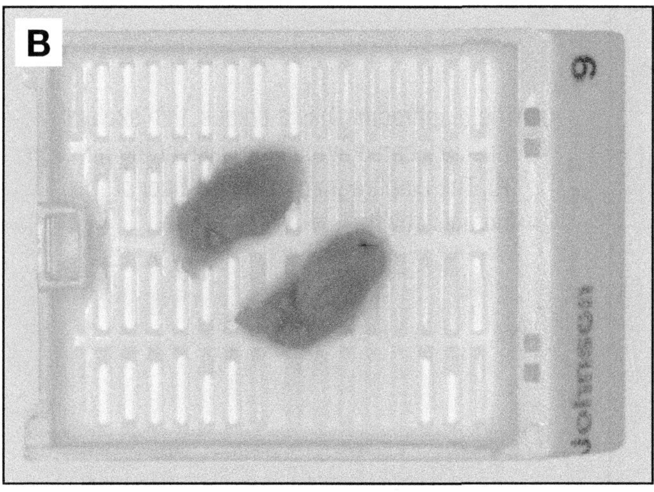

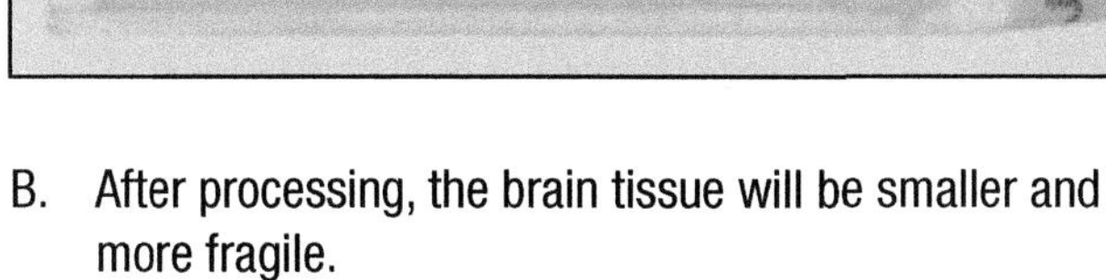

B. After processing, the brain tissue will be smaller and more fragile.

C. Embed both pieces of brain cut side down. Press gently but firmly to ensure the full face is flat on the bottom of the embedding mold.

D. When facing into the block, the goal is to obtain a full face of both pieces of brain.

E. A good section of sagittally trimmed brain should demonstrate cerebrum (CE), cerebellum (CB), hypothalamus (H), brain stem (B) and ventricles (V).

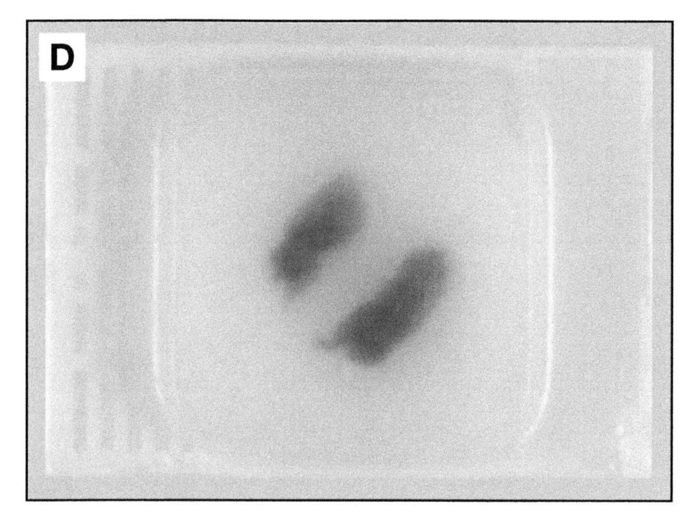

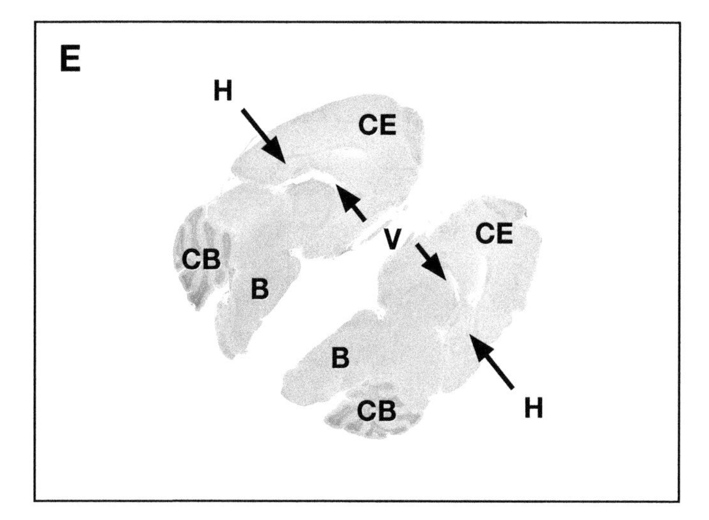

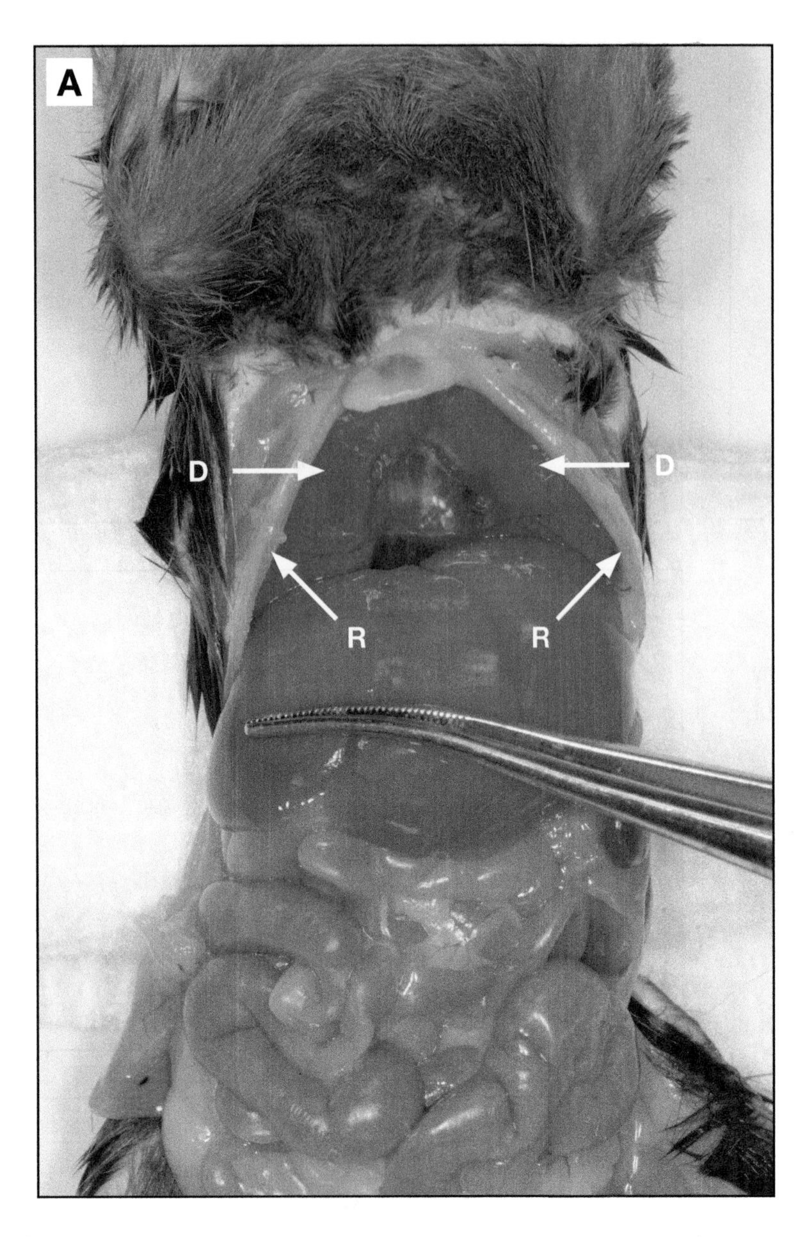

Diaphragm

A. The diaphragm (D) is a unique, thin skeletal muscle located across the base of the rib cage (R). The diaphragm creates the floor of the thoracic cavity.

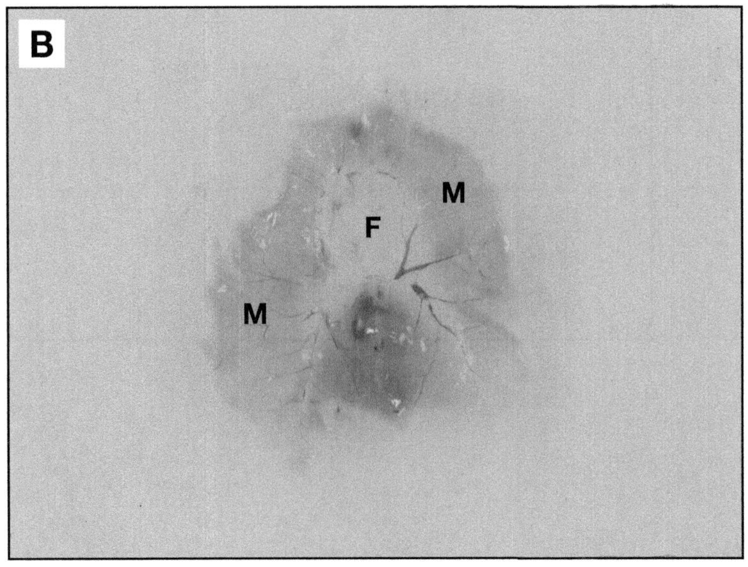

B. Once removed from the mouse, the two different areas of the diaphragm — the muscular area (M) and a clear fibrous area (F) in the center are easier to distinguish. Open the thin diaphragm onto a piece of paper towel prior to cassetting to keep it flat during fixation and processing.

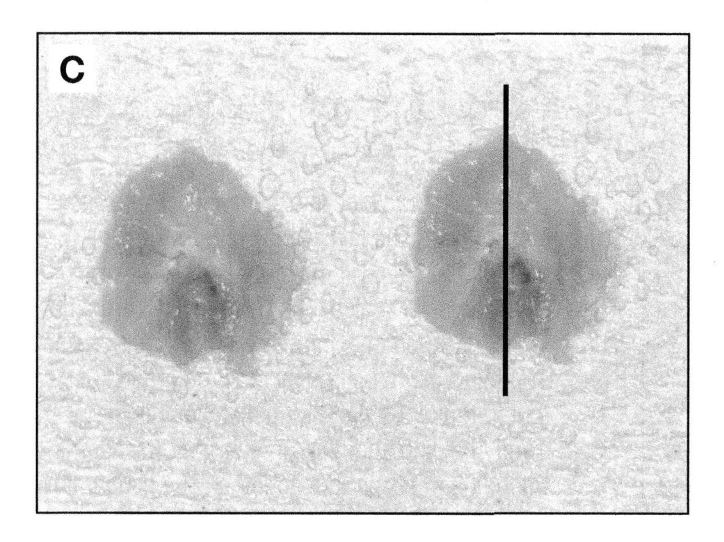

C. After processing, the diaphragm is shrunken and more tan in color. The diaphragm is cut prior to embedding in a sagittal plane from sternum to backbone.

D. To embed the diaphragm, stand these muscles up on the freshly cut edges.

E. The diaphragm will be easier to cut if it is placed on a diagonal as shown in the block. Trim into the block for full faces of both pieces.

F. A good section for analysis contains both the muscular (M) and fibrous (F) areas.

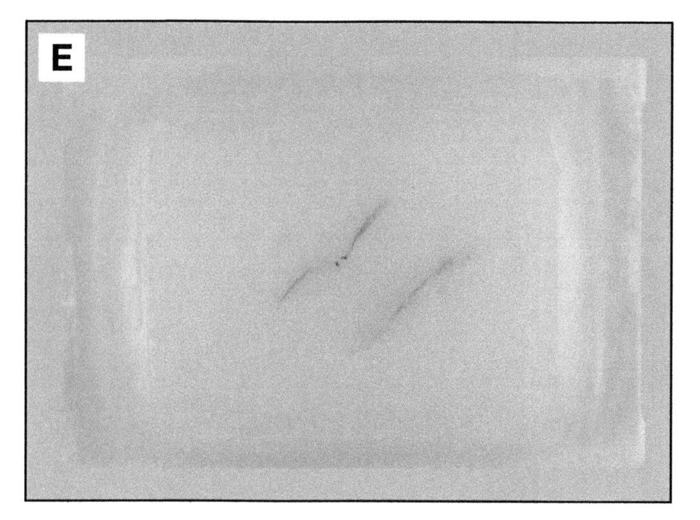

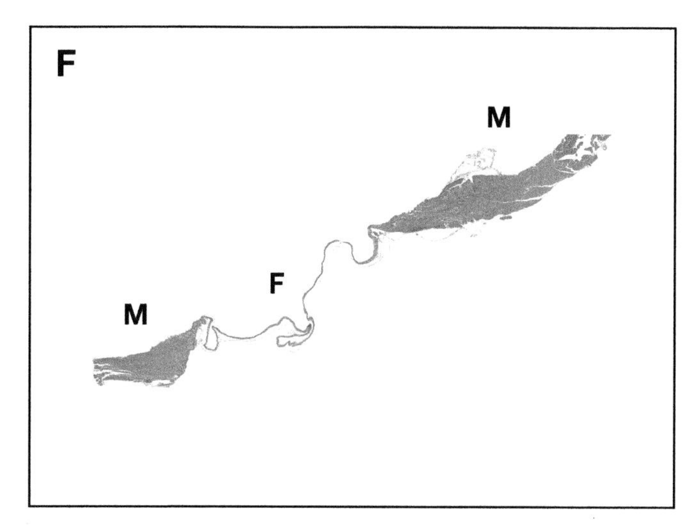

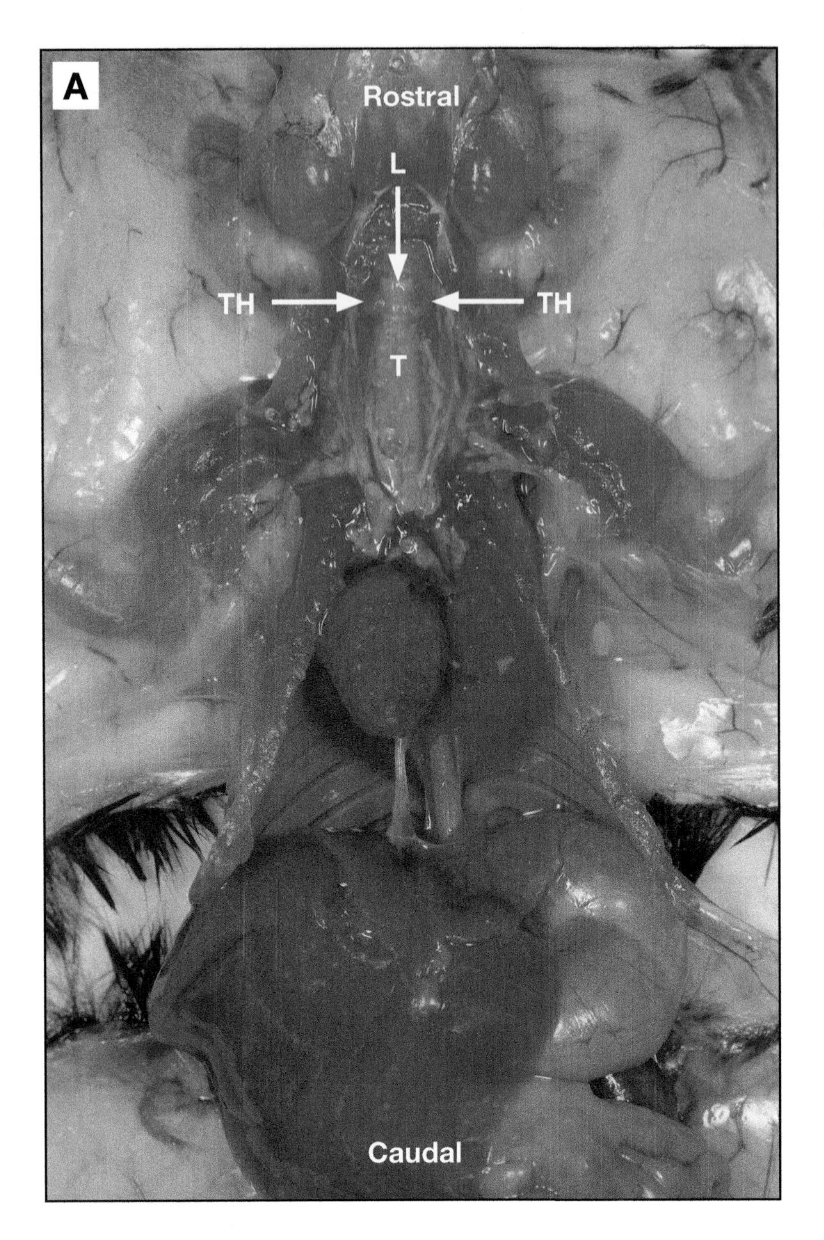

Esophagus, Trachea and Thyroid

A. Once the thoracic cavity is opened, carefully dissect away the layer of muscle over the trachea (T). This will make the larynx (L) and trachea easier to see. The thyroids (TH) are small, dark red in color and visible on either side of the larynx. To collect these three organs as a unit; make the first cut just above the larynx. Gently holding and lifting the larynx, dissect the trachea and esophagus together. The esophagus is attached to the dorsal side of the trachea. Include about 2 mm - 3 mm of trachea and esophagus with the larynx in the cassette for histology.

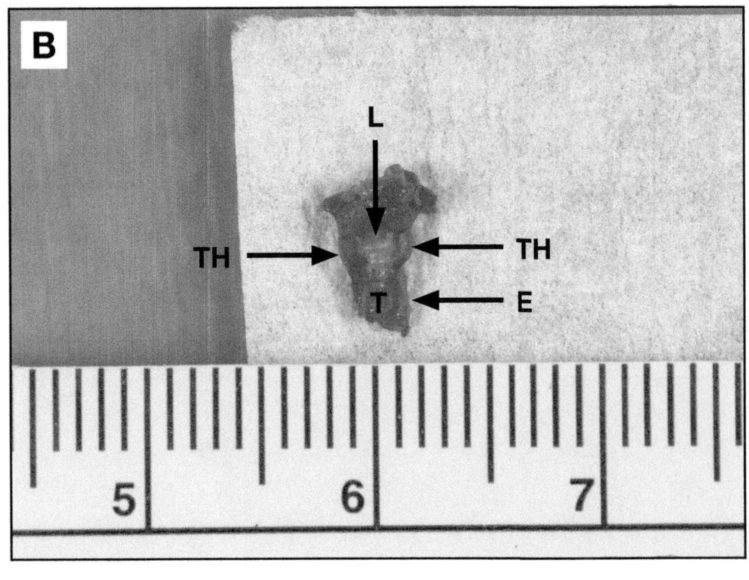

B. Thyroids (TH) and esophagus (E) are easier to see on the larynx (L) and trachea (T) when out of the mouse. Wrap the tissue in a piece of paper towel to help preserve the orientation and integrity of the tissue.

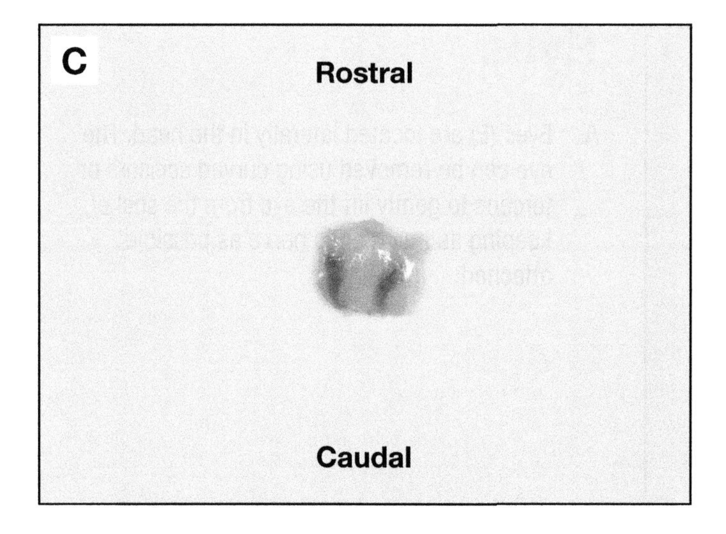

C. After processing, the esophagus, trachea, thyroids and larynx may be harder to distinguish from each other. The larynx at rostral end will be wider than the trachea and esophagus at the caudal end.

D. Stand the specimen with the larynx at rostral end down into the embedding mold. The esophagus and trachea should be facing up.

E. When facing these combined tissues, try to cut deep enough into the specimen to get to the thyroid tissue. This may require frequent checking under the microscope to determine when it has been reached.

F. The section should show trachea (T), thyroid (TH) and esophagus (E). Occasionally, cervical lymph nodes (LN) may also be present in these sections.

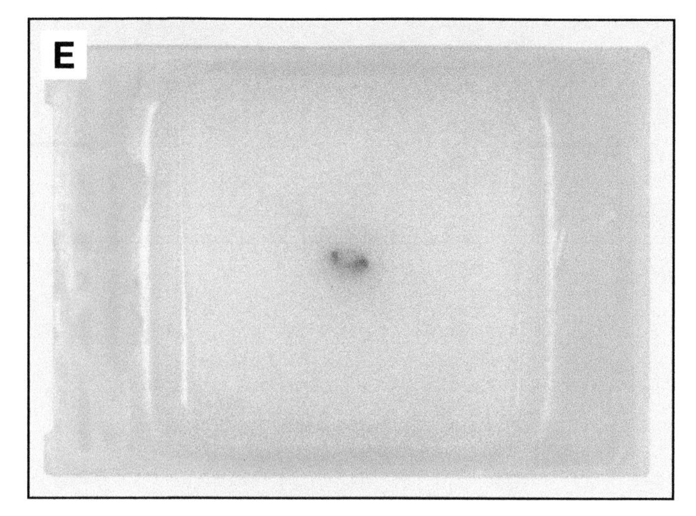

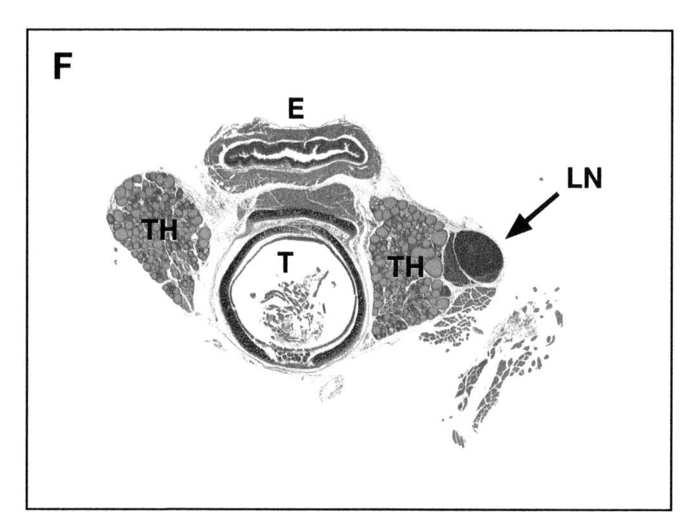

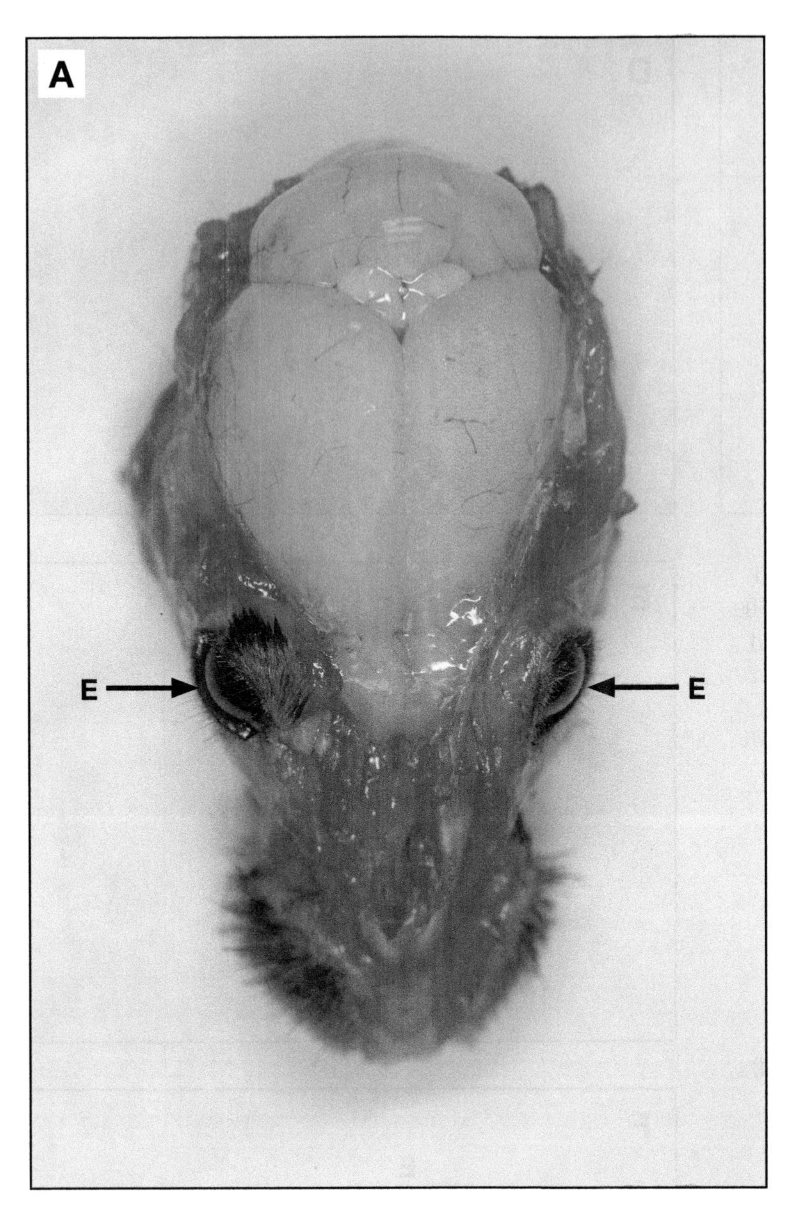

Eyes

A. Eyes (E) are located laterally in the head. The eye can be removed using curved scissors or forceps to gently lift the eye from the socket, keeping as much optic nerve as possible attached.

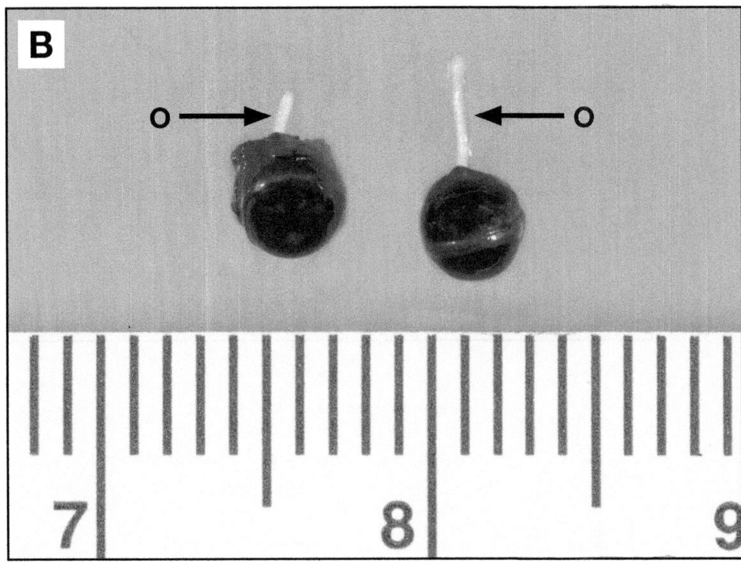

B. It is recommended that the eyes be removed and placed into in Davidson's fixative to preserve the rods and cones as well as the nuclear layers of the retina. After 24 hours the eyes should be transferred to neutral buffered formalin (NBF). Optic nerves (O) can be seen here.

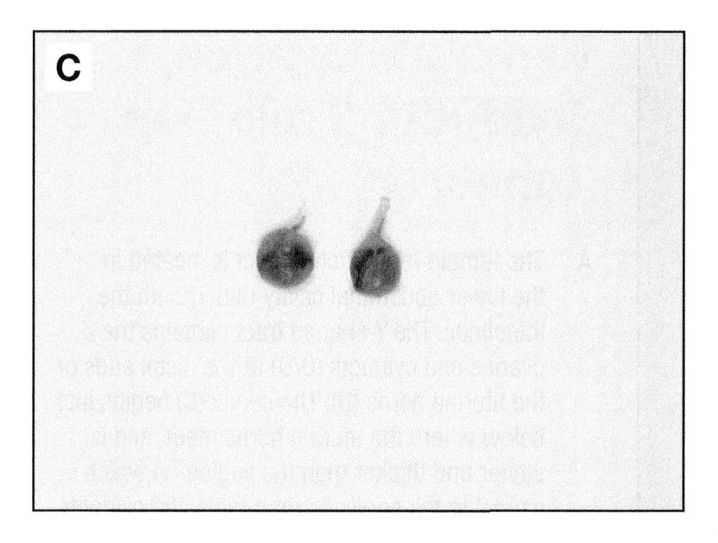

C. After processing, the eyes are shrunken and may have lost some of their roundness.

D. To embed, carefully place the eyes into the embedding mold with the optic nerve parallel to the cutting surface.

E. Face into the block enough to obtain a section with cornea and, if possible, optic nerve.

F. A section of eye should contain cornea (C), lens (L), retina (R) and optic nerve (O) from at least one and ideally both eyes. This eye (F) was fixed first in Davidson's followed by NBF. There is better nuclear preservation in the retina.

G. This section (G) was taken from an eye fixed in NBF only for comparison. There can be separation of the retina, a common artifact from NBF fixation of the eye.

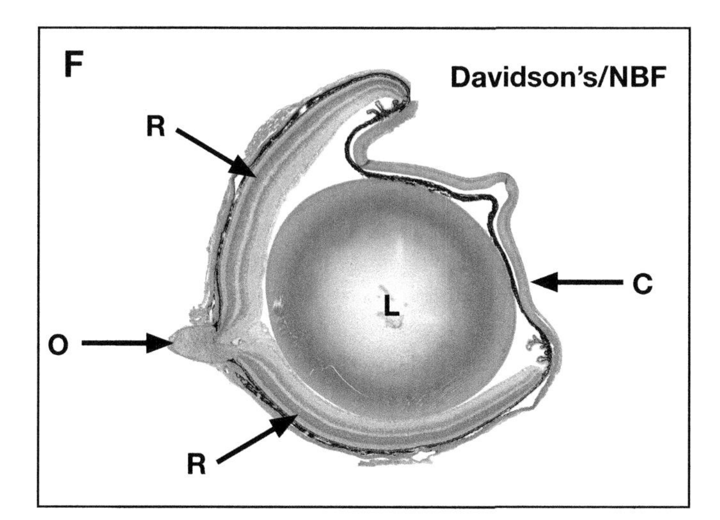

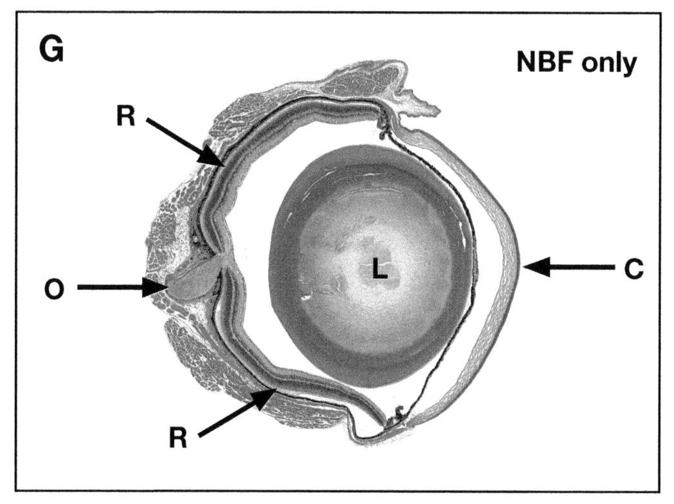

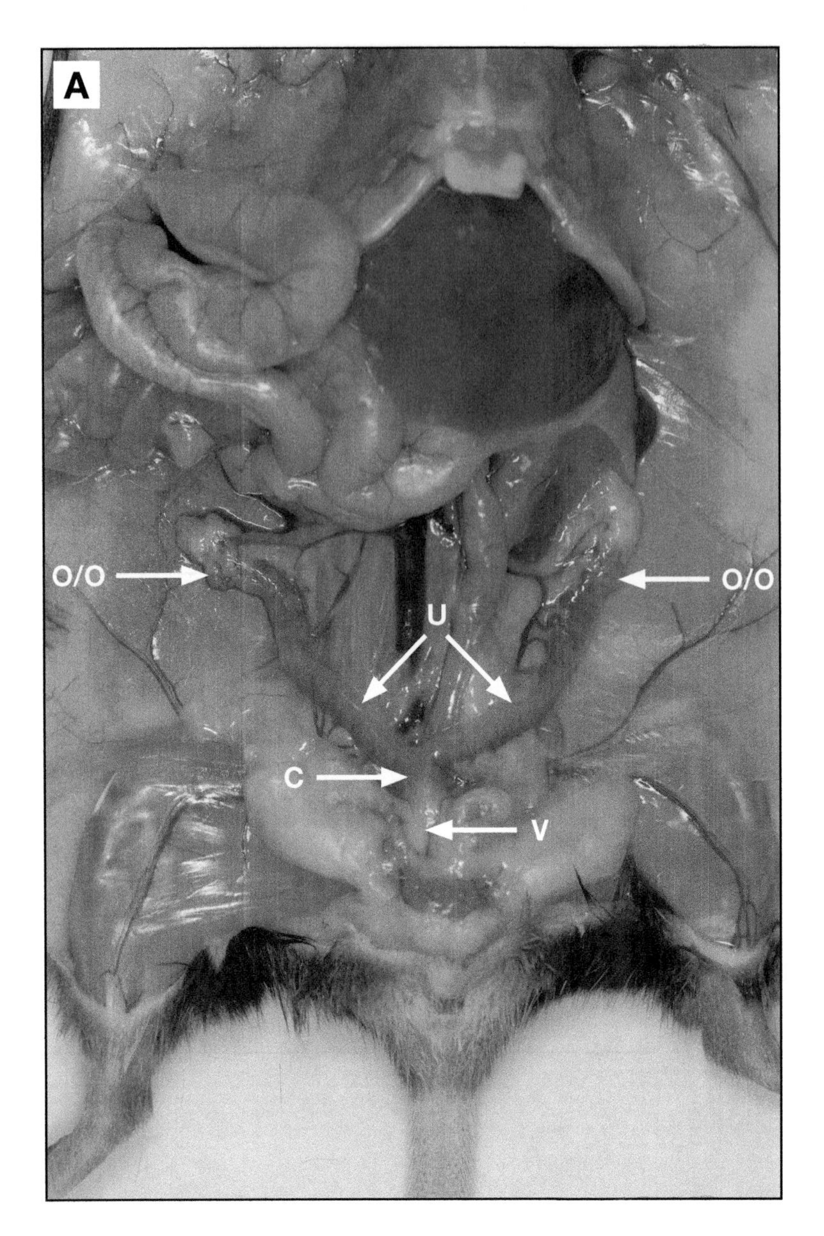

Female - Ovaries, Oviducts (Fallopian Tubes)

A. The female reproductive tract is located in the lower abdominal cavity underneath the intestines. The Y-shaped tract contains the ovaries and oviducts (O/O) at the distal ends of the uterine horns (U). The cervix (C) begins just below where the uterine horns meet, and is whiter and thicker than the vagina (V) which is caudal to the cervix. In mammals, the oviducts are sometimes referred to as the Fallopian tubes.

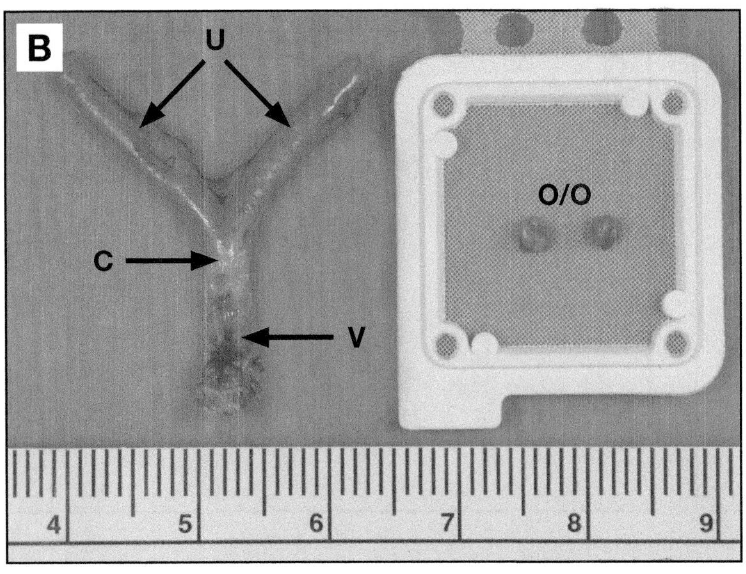

B. In this photo, the ovaries and oviducts (O/O) are removed from the uterine horns (U) and either placed into a CellSafe™ Biopsy Insert or wrapped in paper towel to prevent their loss during processing. The ovaries and oviducts are almost impossible to distinguish from each other without the aid of a microscope. Cervix (C) and vagina (V) are easier to identify when removed from the body.

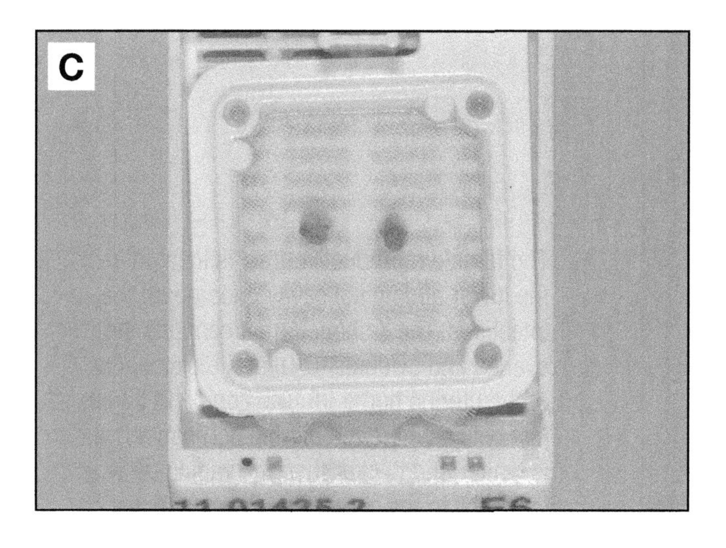

C. After fixation and processing, the ovaries and oviducts are smaller and darker in color.

D. When embedding the ovaries, ensure they are as flat as possible in the bottom of the mold.

E. If possible, face into the block until the wide faces of both ovaries are visible. Use care as these are very small tissues.

F. The ideal section will contain ovary (O) and oviduct (OD).

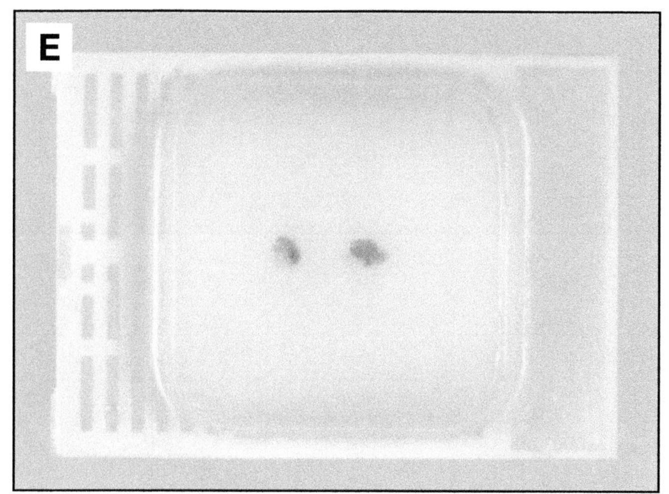

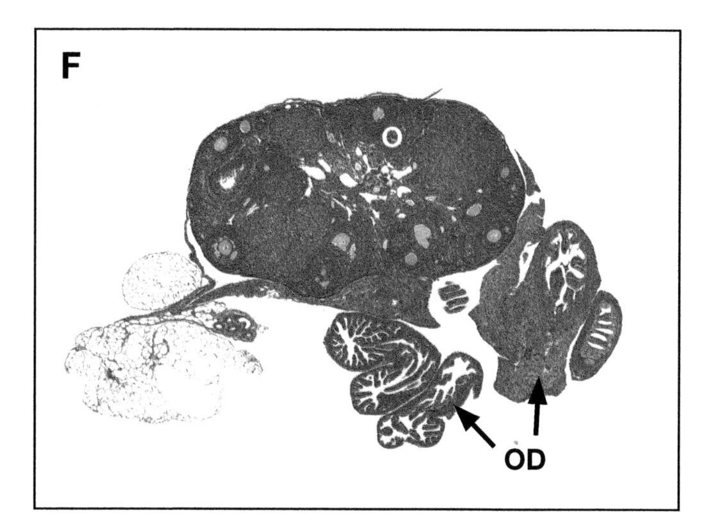

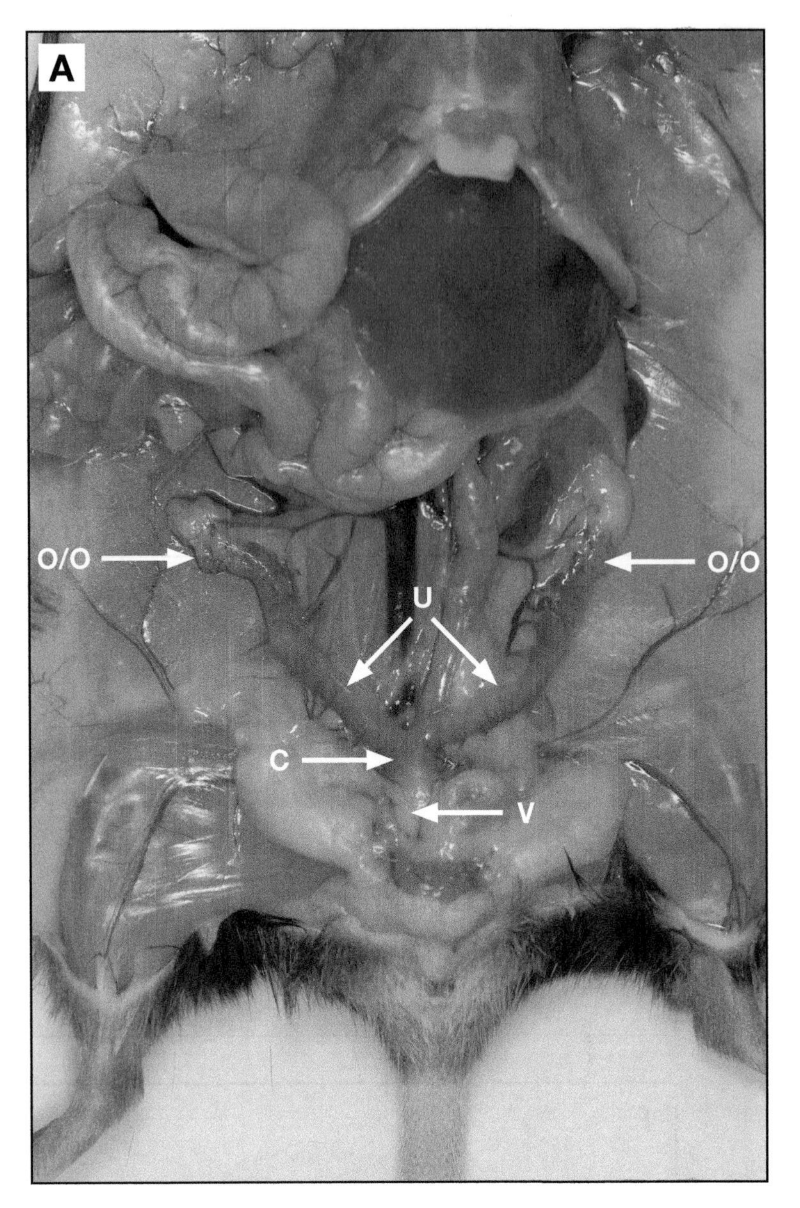

Female - Uterus (Uterine Horn), Cervix, Vagina

A. The female reproductive tract is located in the lower abdominal cavity underneath the intestines. The Y-shaped tract contains the ovaries and oviducts (O/O) at the distal ends of the uterine horns (U). The cervix (C) begins just below where the uterine horns meet, and is whiter and thicker than the vagina (V) which is caudal to the cervix. The cervix and vagina sit under the pelvic bone which must be cut in order for the reproductive tract to be removed.

B. The ovaries and oviducts (O/O) are removed from the uterine horns and cassetted separately. The remaining Y contains the uterine horns (U), cervix (C) and vagina (V). The remaining "Y" shaped tissues are laid flat and wrapped in a piece of paper towel to hold them outstretched during fixation.

C. Trimming for histology can be done at any point before embedding but is best done after some fixation. Here, two transverse sections are made through the horns of the uterus (lines). The caudal end of the vagina is trimmed off and discarded. The smaller, now trimmed Y (uterus, cervix and vagina) and the transverse sections of uterus are processed.

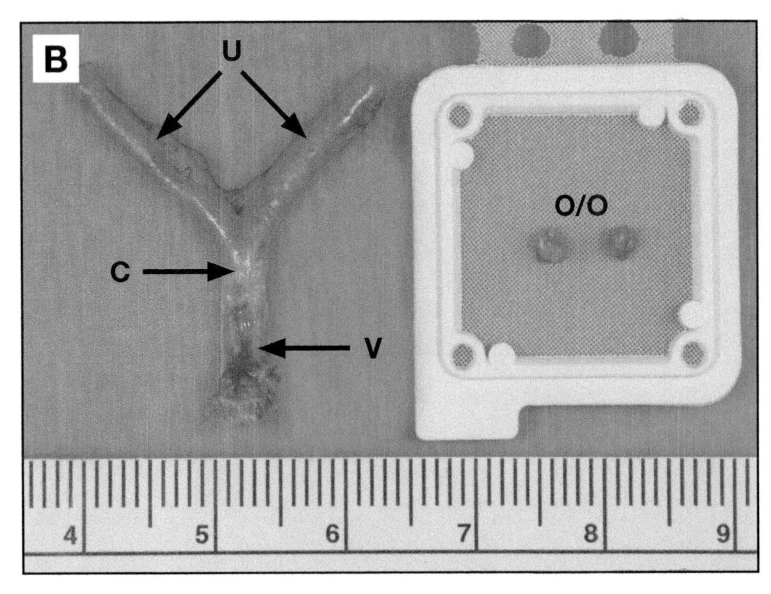

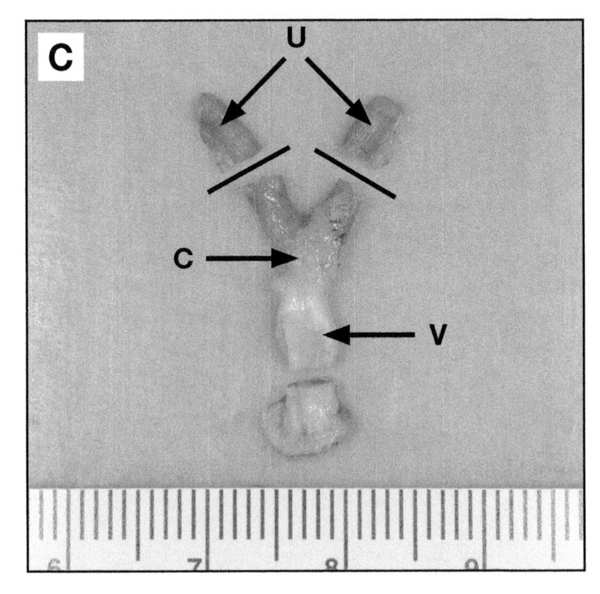

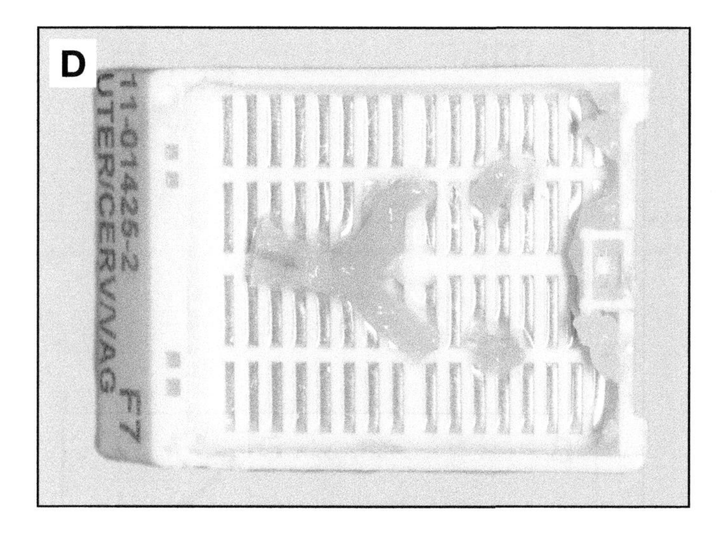

D. After processing, all pieces of tissue are smaller and whiter in color.

E. The trimmed tubes of uterine horn are embedded on end for cross sections. The remaining reproductive tract is placed flat into the mold, creating the “Y” shape.

F. Face into the block until lumens of both the vagina and uterine horns are reached.

G. The ideal section contains a complete section through the uterine horns (U) and a clear longitudinal section through the cervix (C) and vagina (V).

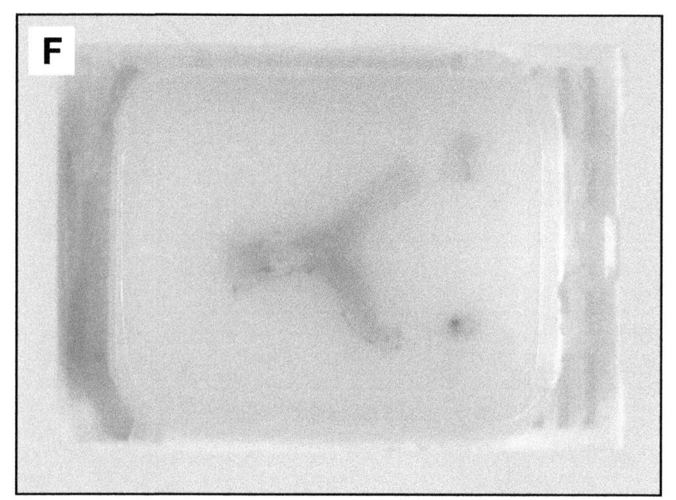

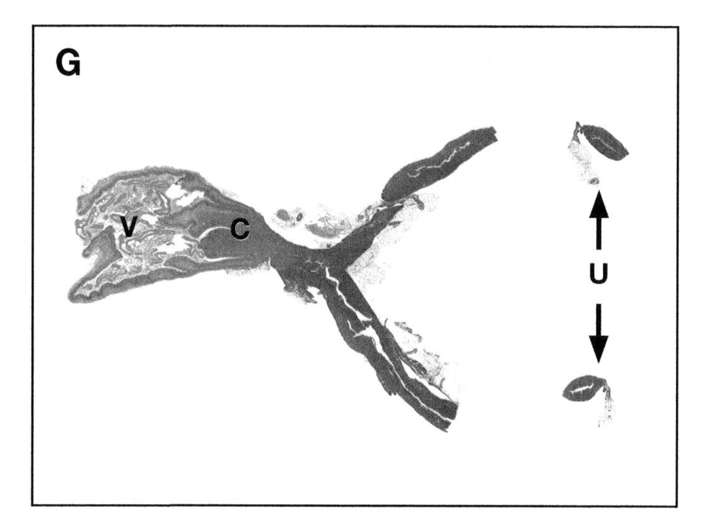

Femur

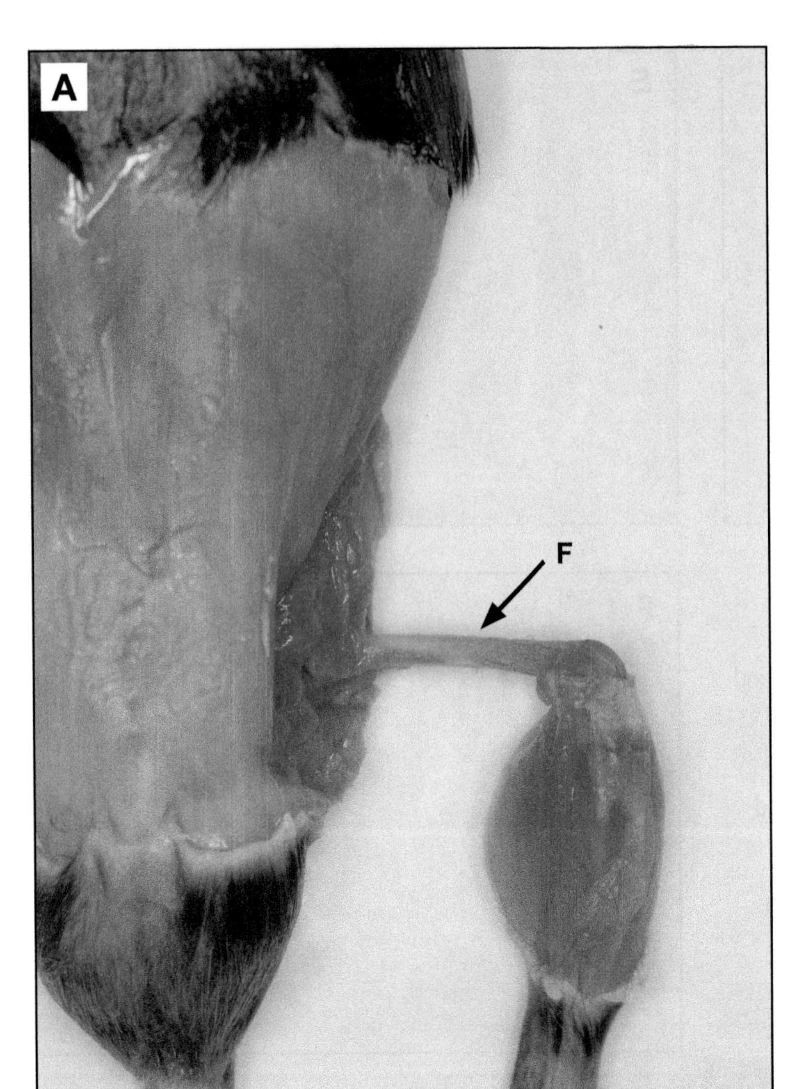

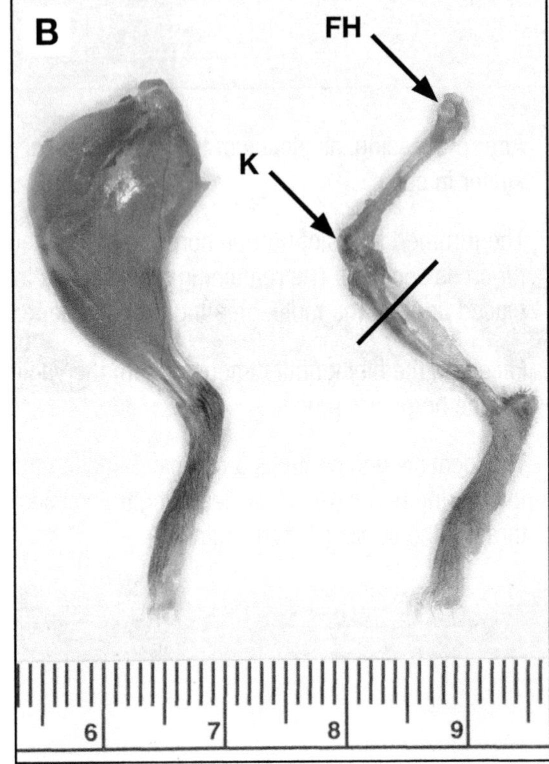

A. In this photograph, the skin and muscle have been removed from the thigh to expose the femur (F).

B. Remove the femur from the hip joint being careful to collect the femoral head (FH) intact. Removing as much muscle as possible improves the quality of fixation and decalcification. The lower leg bones and foot should be trimmed off at the line just below the knee (K).

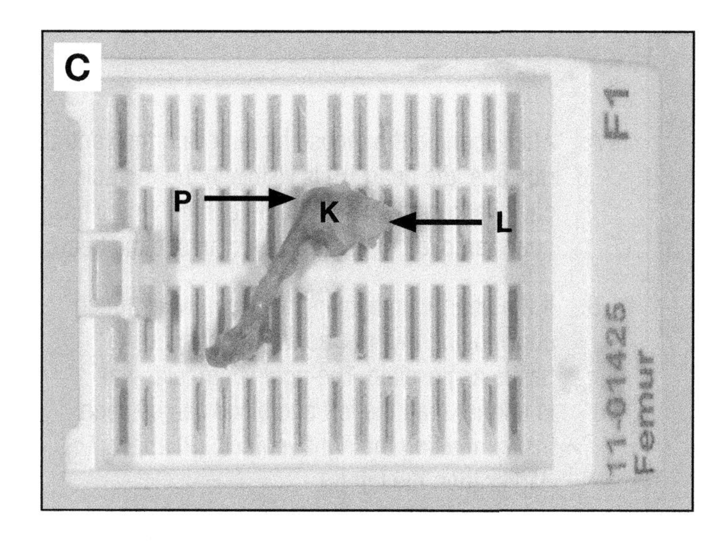

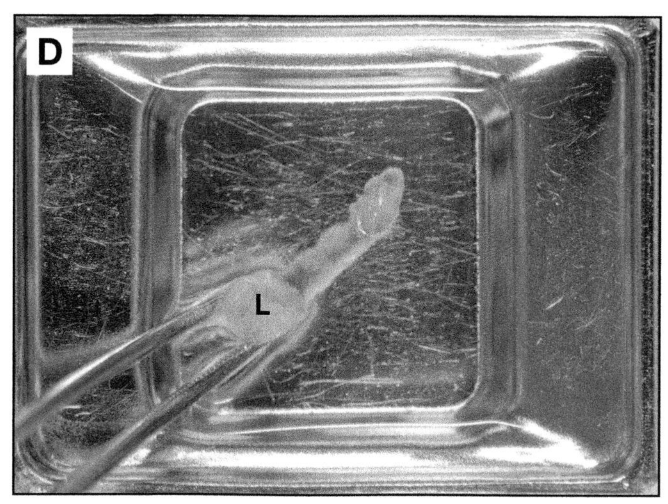

C. After fixation and processing, the muscle has lost its red color and appears more tan. The proximal end of the femur (P), knee (K) and remaining stub of the lower leg (L) are marked on this processed femur.

D. To embed the femur, hold the remaining stub of the lower leg (L) with the forceps and place the opposite side facing down into the embedding mold as flat and level as possible.

E. To examine growth plate in the distal end, (labeled G in plate F below), it is important to have the ends of the femur flat. To examine bone marrow, trim far enough into the specimen to reach the shaft (S) of the bone. In the paraffin block, the marrow will appear darker than the bone.

F. The ideal section should contain growth plate (G), articular cartilage (C) and bone marrow (BM).

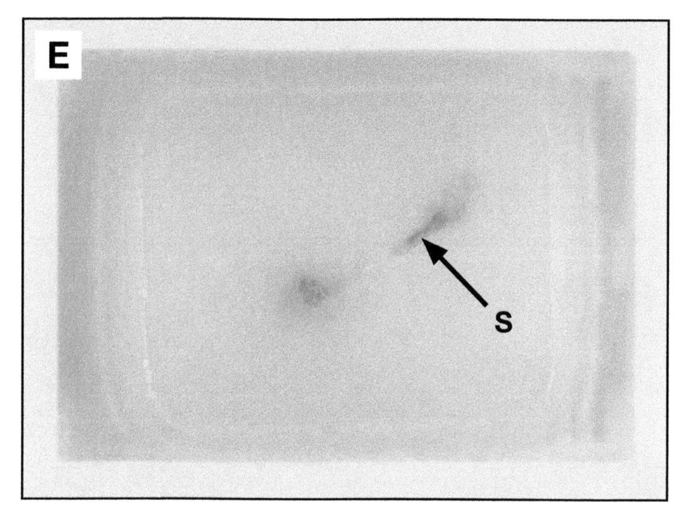

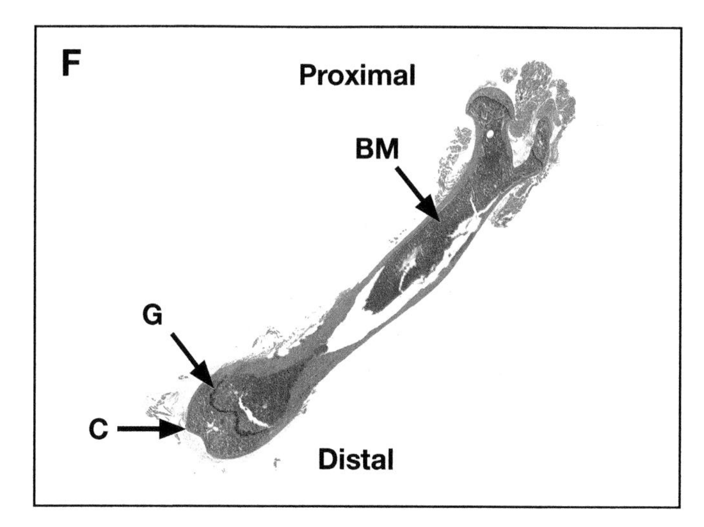

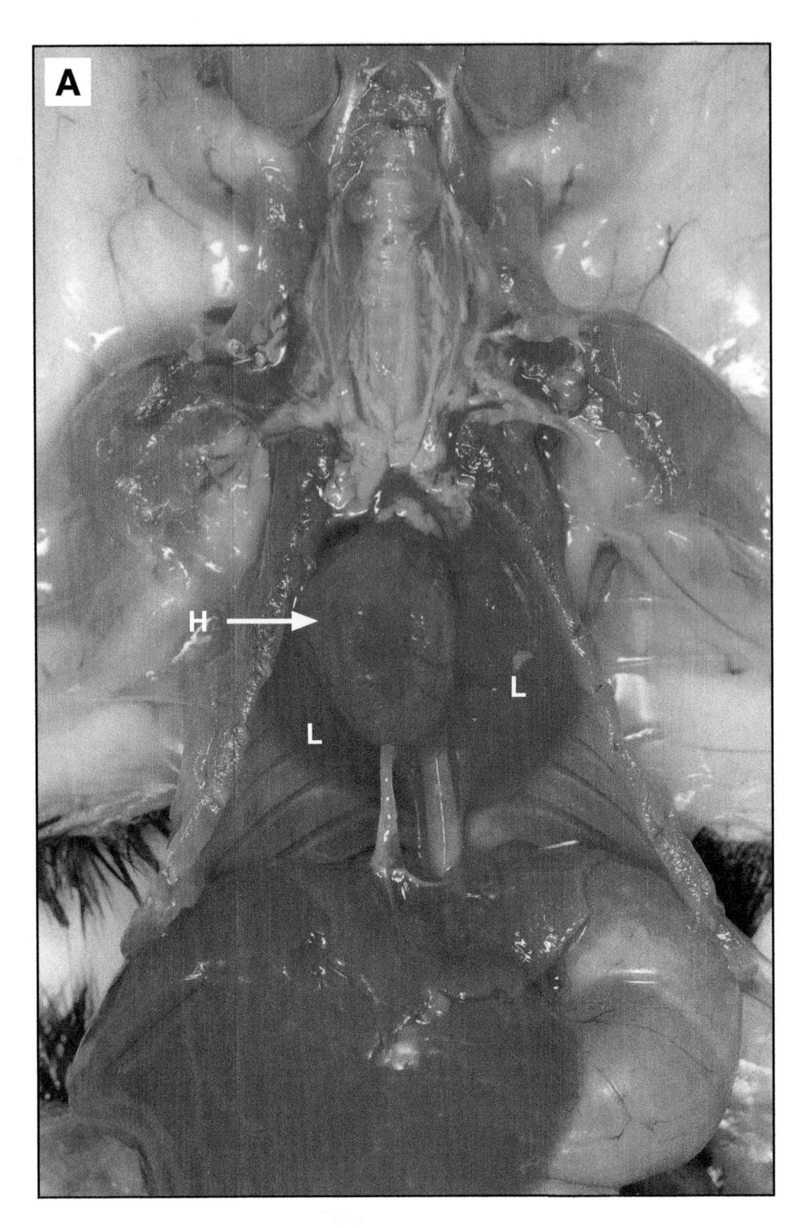

Heart

A. After opening the mouse by removing the ventral rib cage, the heart (H) is visible in the thoracic cavity above the lungs (L). Here, the thymus has been removed to make it easier to see the heart.

B. If the heart is to be weighed, remove as much surrounding thymus, aorta, and connective tissue as possible. The right ventricular wall is thinner than the left and blood can be seen inside, making the right ventricle (R) appear darker than the left (L).

C. After fixation, the heart is trimmed through the apex to reveal the 4 chambers as follows:

1. Locate the highest point of the right ventricle between the two auricles and face this point upward.
2. Section the heart longitudinally through this high point and continue through the apex.
3. After trimming a small piece of paper towel should be used to hold the tissue flat during processing.

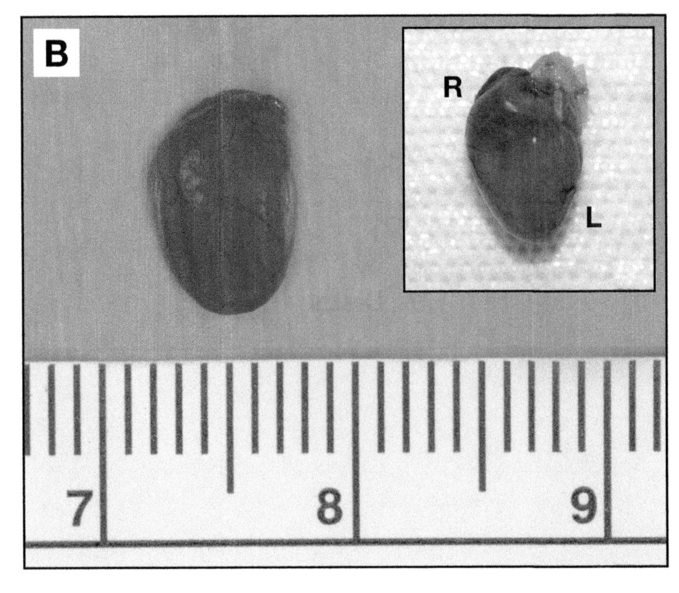

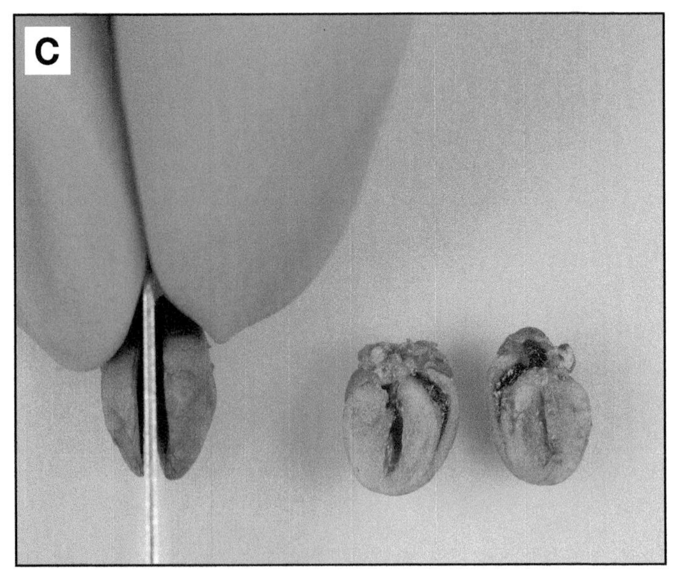

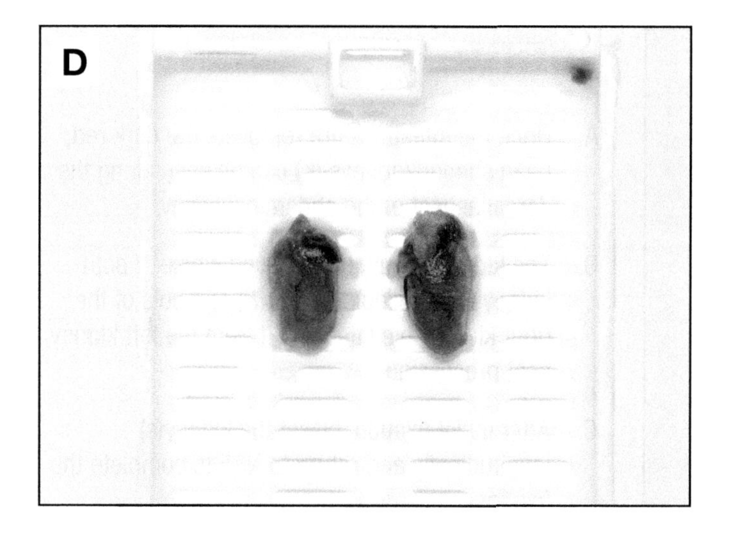

D. After processing, the heart is smaller and has lost its bright red color.

E. Embed the two halves of the heart cut side down after processing.

F. Trim into the block until a full face of each piece of the heart is visible.

G. A section for pathological analysis should have all 4 chambers of the heart and some of the valves visible as well.

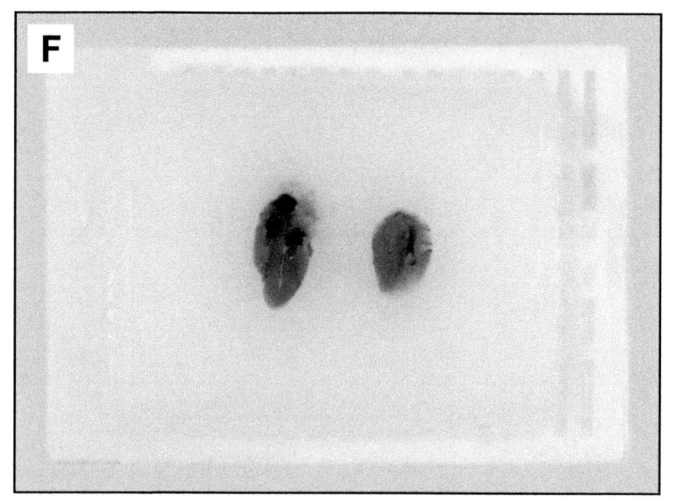

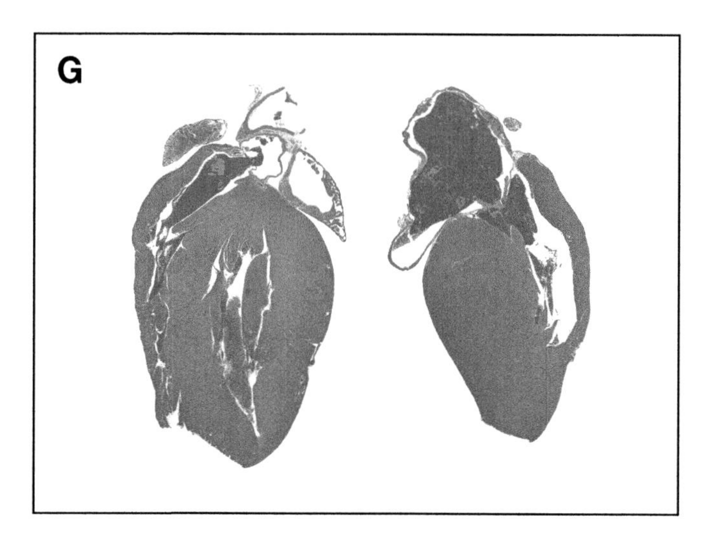

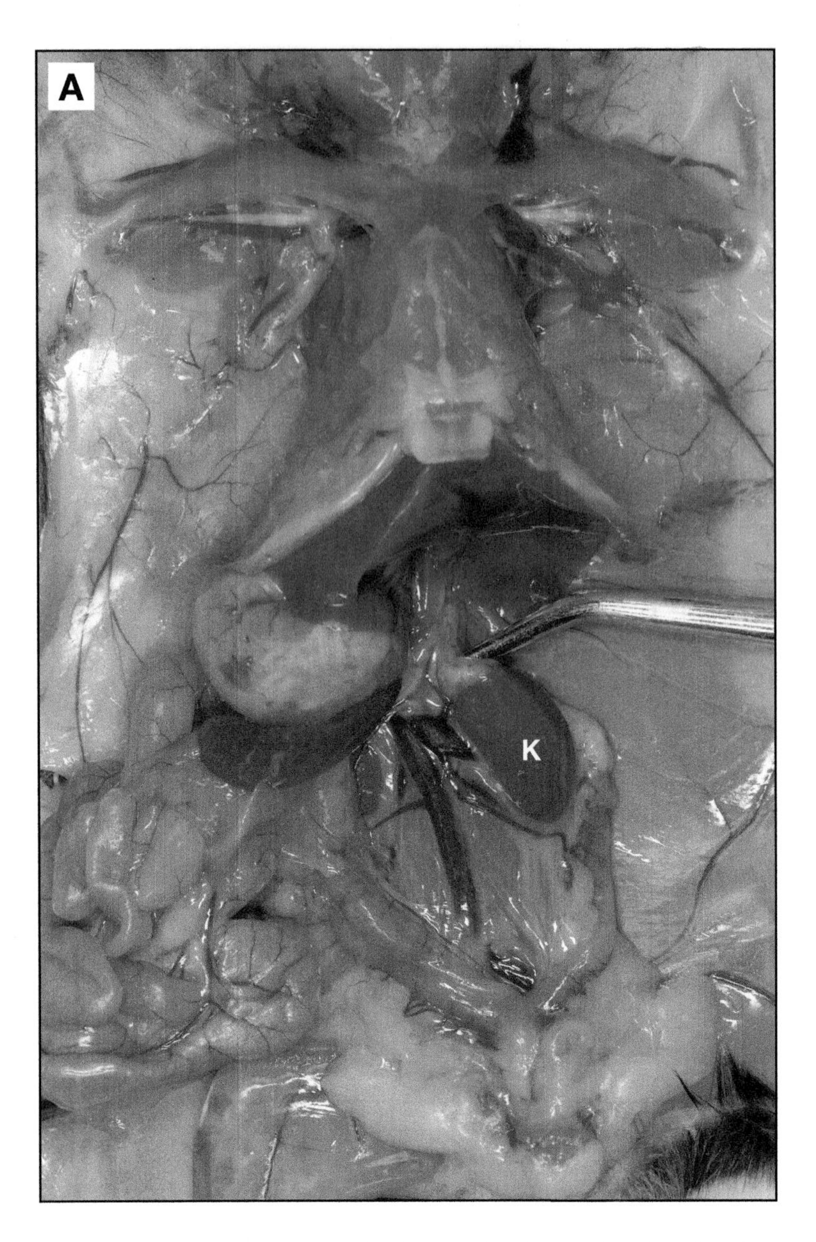

Kidneys

A. Once the intestines are removed, the dark red, bean shaped kidneys (K) can be seen along the dorsal aspect of the abdominal cavity.

B. The kidney should be removed whole. If both kidneys are collected, nick (N) one pole of the right kidney to distinguish it from the left kidney, post processing.

C. After initial fixation, bisect the kidney(s) longitudinally and return to NBF to complete the fixation.

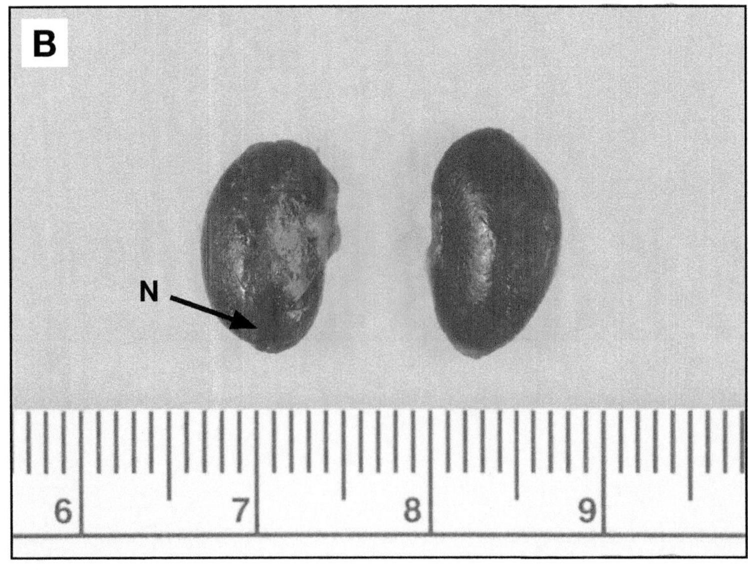

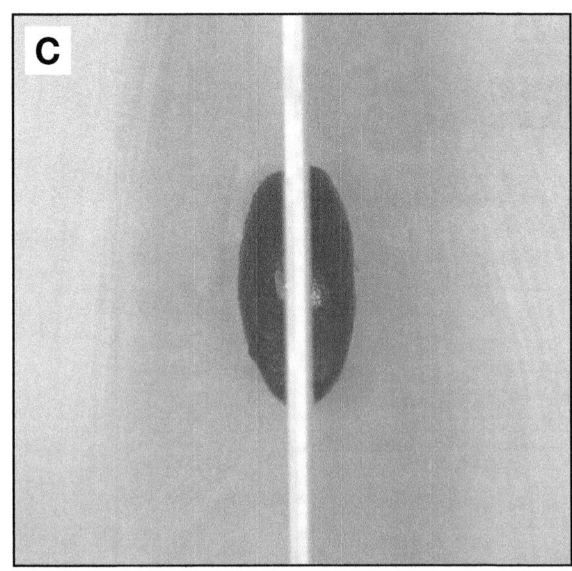

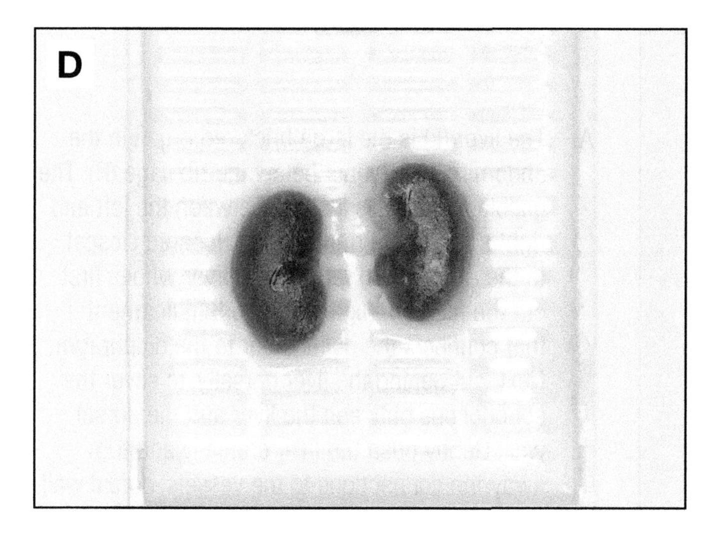

D. After processing, the kidney is smaller and darker in color.

E. Embed the kidneys cut face down into the embedding mold.

F. Face into the block until the whole surface of each kidney is exposed.

G. The best section should have at least one complete longitudinal kidney section with both cortex (C) and medulla (M).

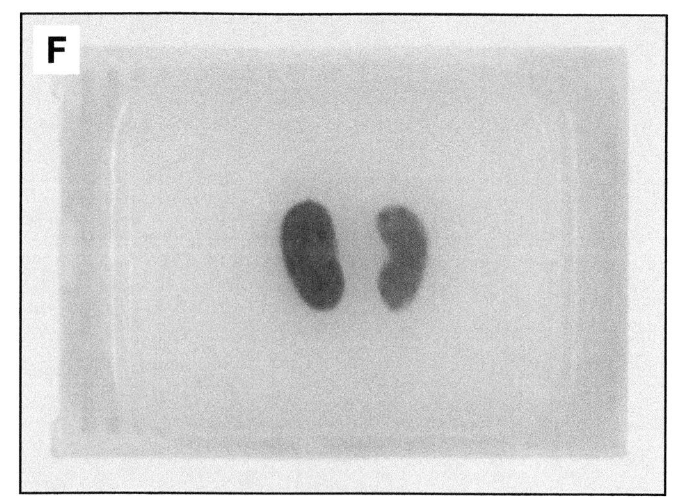

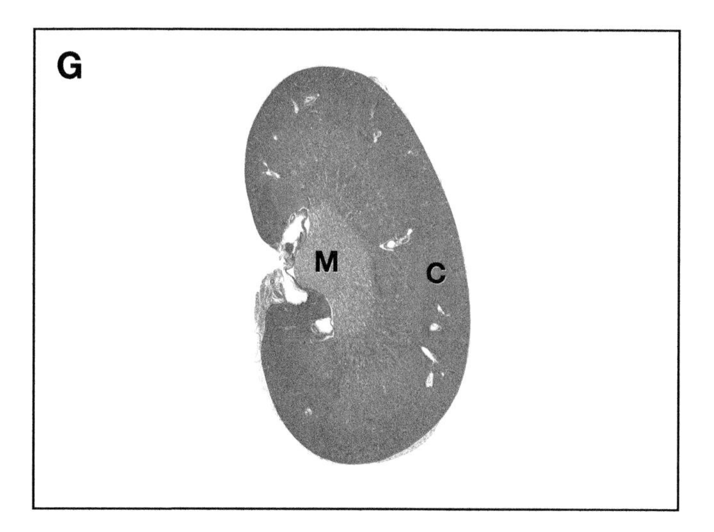

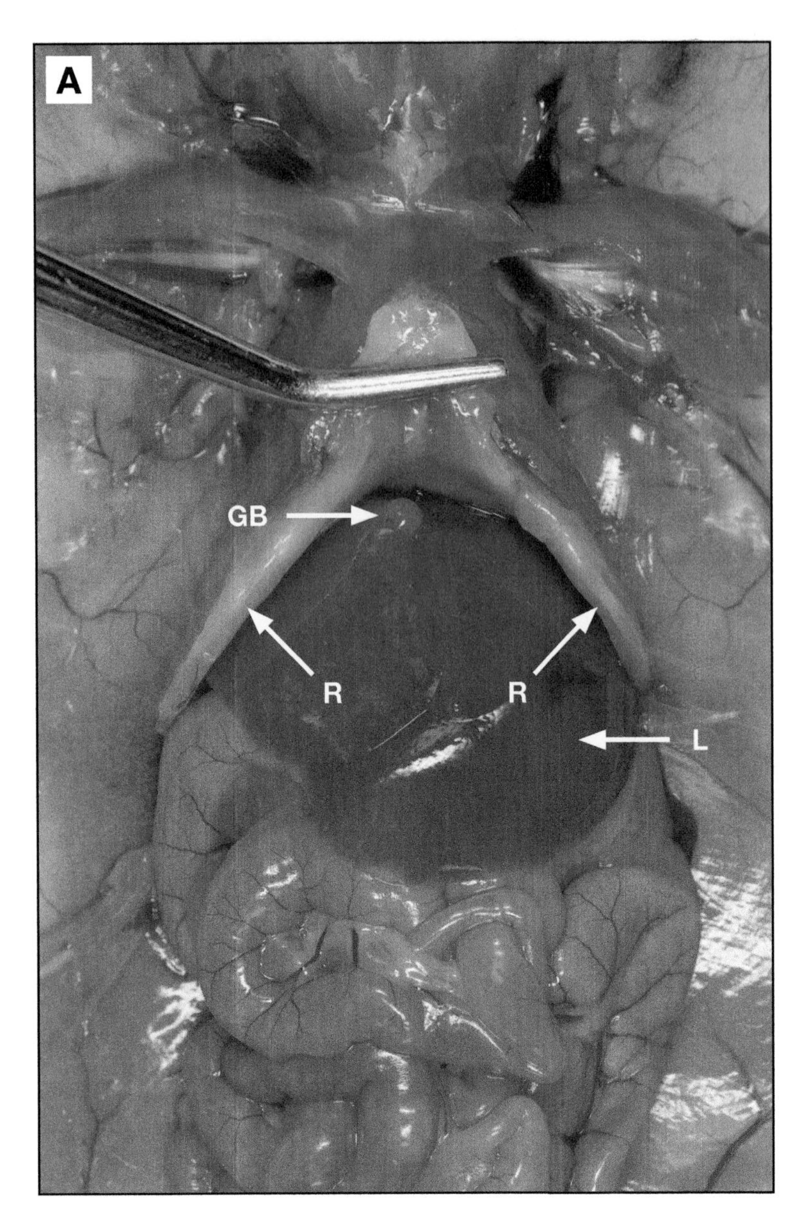

Liver and Gallbladder

A. The liver (L) is the large brick red organ in the abdominal cavity just below the rib cage (R). The gallbladder (GB) is located between the left and right sides of the median lobe, located closest to the rib cage. To remove the liver whole, first cut the falciform ligament, the thin ligament that connects the gallbladder to the diaphragm. Next, cut behind the liver dorsally, to sever the connections between the liver and the dorsal wall. Gently push the liver cranially and trim away the connections to the vessels, dorsal wall and stomach. The right adrenal gland is closely associated with the lower lobes of the liver, therefore care is needed to locate the adrenal gland (see page 2) prior to removal of the liver. When the entire liver is required for weighing, be sure to locate all lobes, especially small lobes attached to the stomach and near the intestines.

B. The left lateral lobe is the largest lobe of the liver. It lies immediately below the median lobe which contains the gallbladder. Once the lateral lobe is separated from the rest of the liver, collect a piece 3 mm - 5 mm in width by trimming from the hilium, where the liver connects to the vessels dorsally, to the edge of the liver.

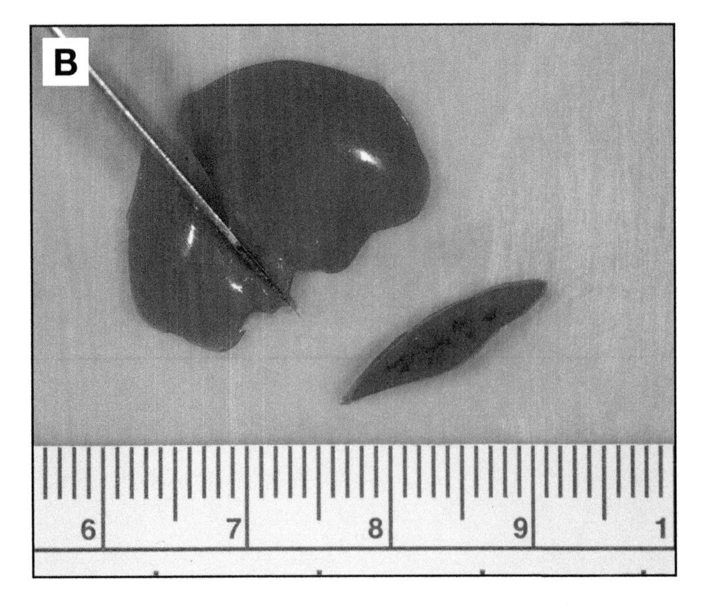

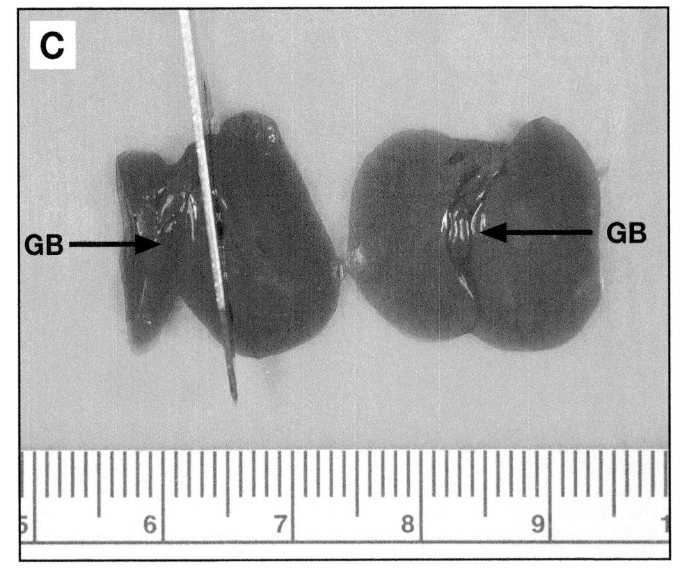

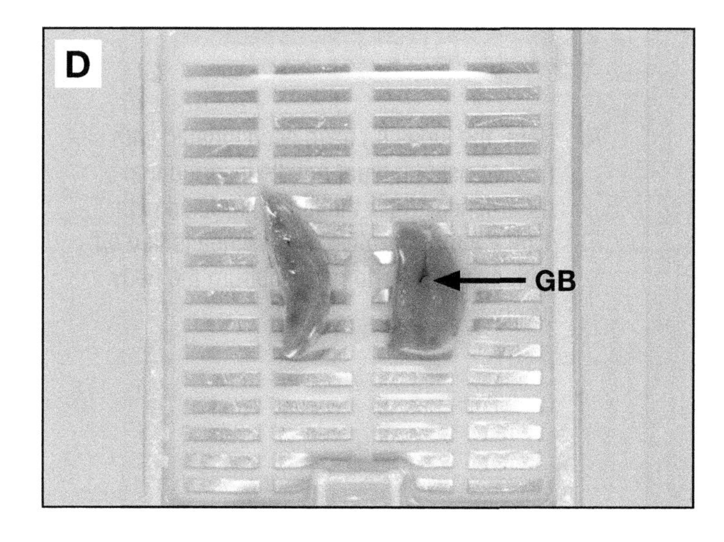

C. To sample the gallbladder (GB), collect and remove the median lobe from the whole liver. This is the most cranial lobe that is located directly below the ribs. Trim into the liver from the hilium, where the liver connects to the vessels, to the edge of the liver. This is about 2 mm - 3 mm to the left and right of the gallbladder as shown in photo.

D. The liver will become less red and more brown to grey in color and the gallbladder (GB) may become smaller and harder to see after fixation and processing.

E. At embedding, place the sample from the left lateral lobe with either cut face down. When embedding the median lobe, first locate the gallbladder. Place the side where the gallbladder is easiest to see face down into the mold.

F. Trim into the block until the gallbladder is visible and a full face of the left lateral lobe can be seen.

G. A perfect section will contain the complete gallbladder (GB) in the median lobe and a full face of the left lateral lobe containing parenchyma (P) and portal triads (T).

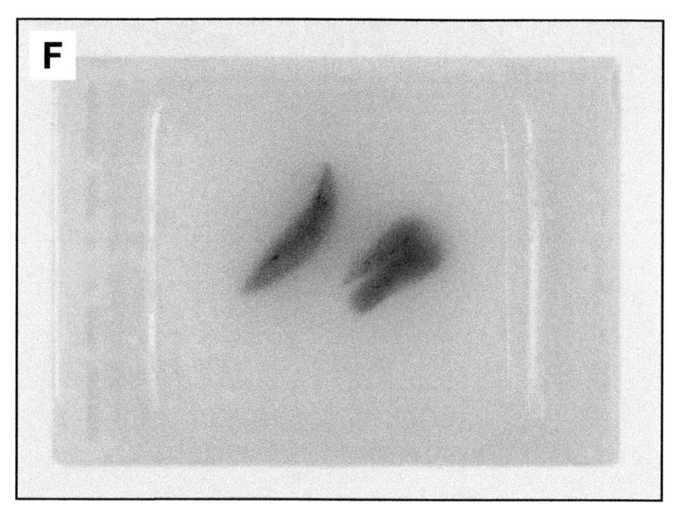

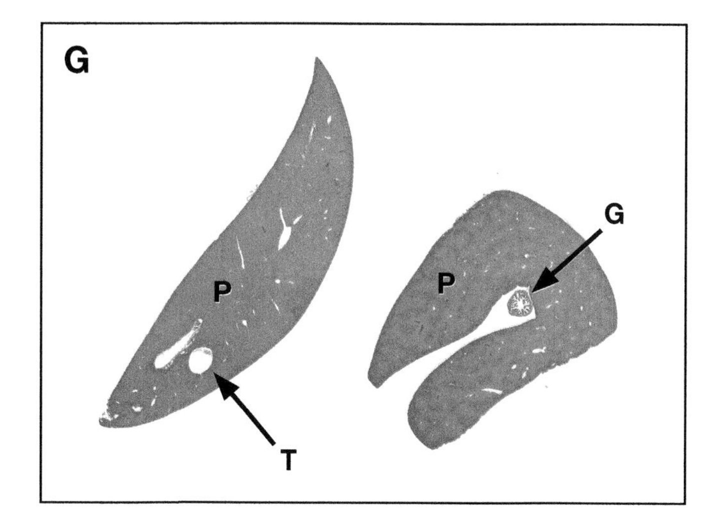

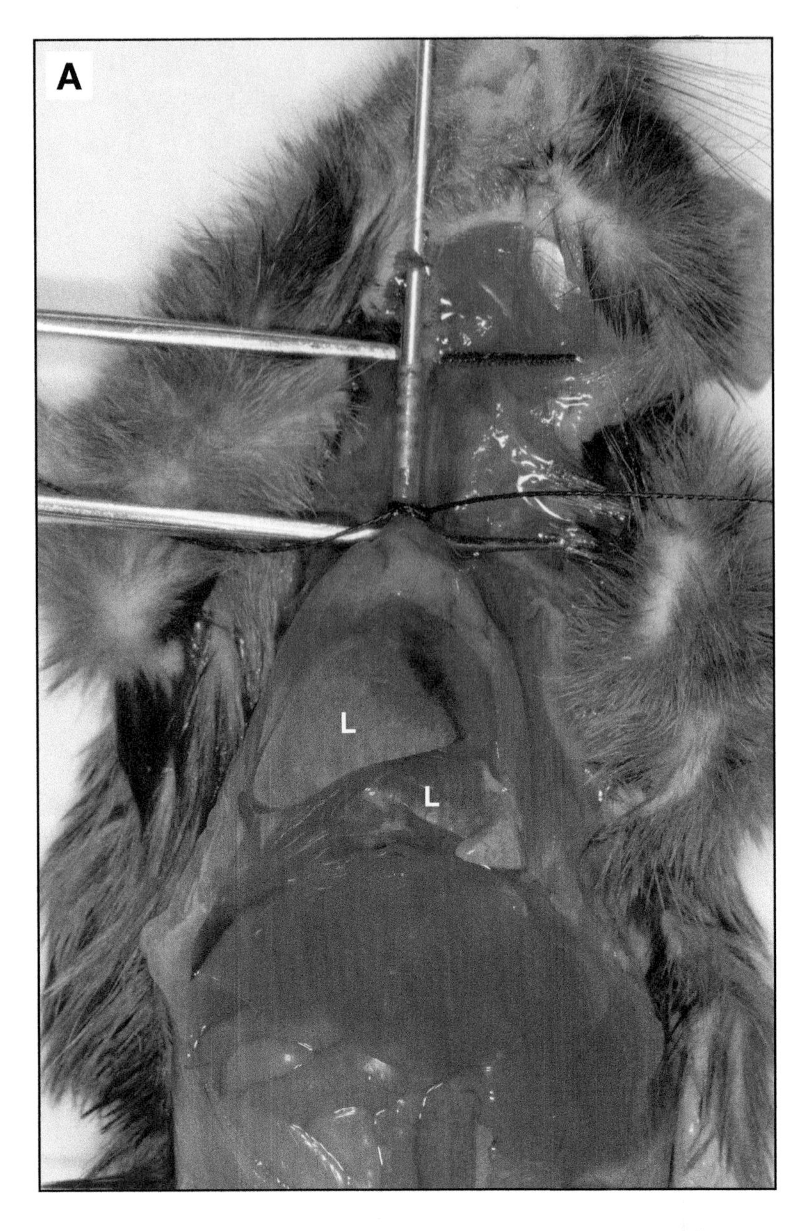

Lung (Inflated)

A. Here, the lungs (L) have been inflated with fixative using a blunt needle inserted into the trachea. After the lungs are filled with 1 ml - 2 ml of fixative, the trachea is tied off with suture prior to removal from the body cavity.

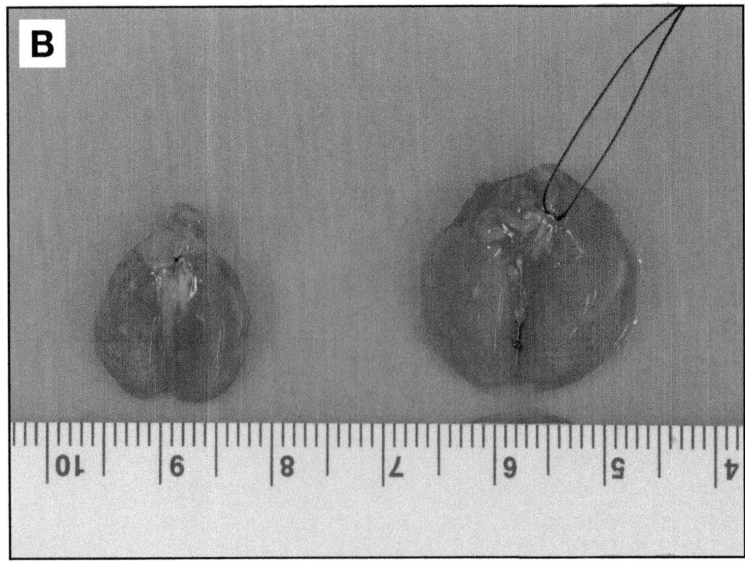

B. A comparison of uninflated (left) and lungs inflated with fixative (right). Inflating the lungs with fixative opens the alveolar spaces allowing for better preservation than immersion fixation alone.

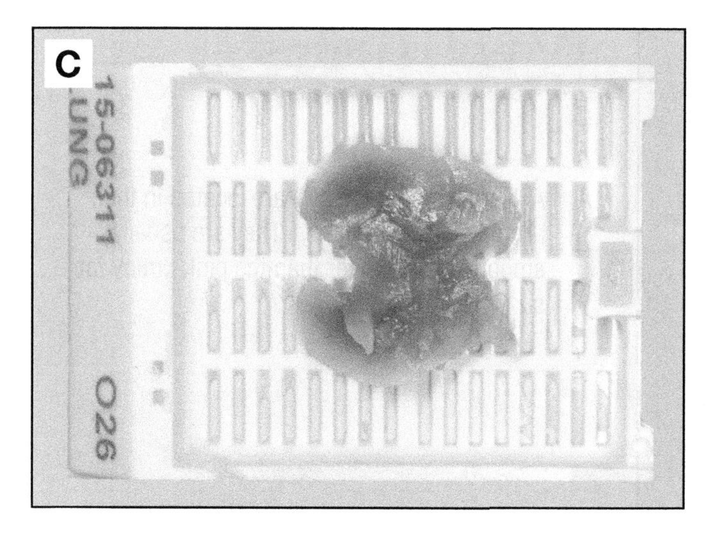

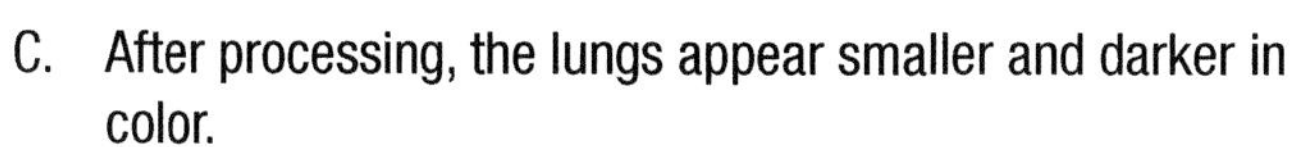

C. After processing, the lungs appear smaller and darker in color.

D. To embed, place the lungs with the rounded edges, dorsal face, down into the mold. Try to spread out the lobes as they are pushed down.

E. When facing into the block, try to get into as many lobes as possible (minimum of 3).

F. The ideal section will contain at least three separate lobes. In addition to the parenchyma (P), main bronchi (B) and major blood vessels (V) should be visible. The opening of the alveolar spaces with fixative allows for better evaluation of histological changes, especially in the peripheral parenchyma.

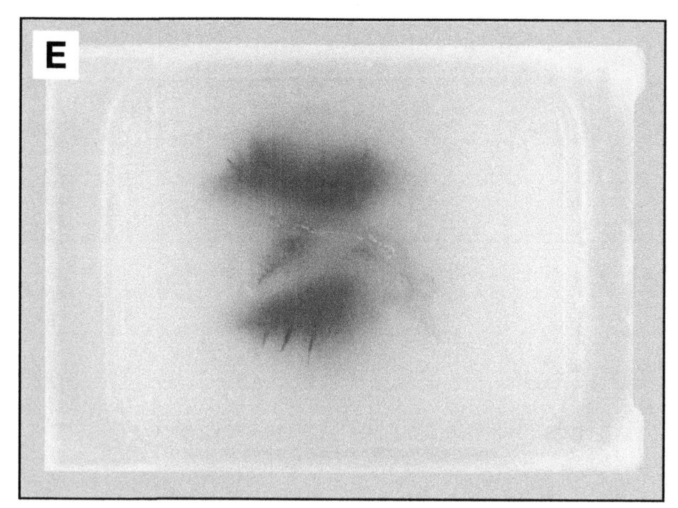

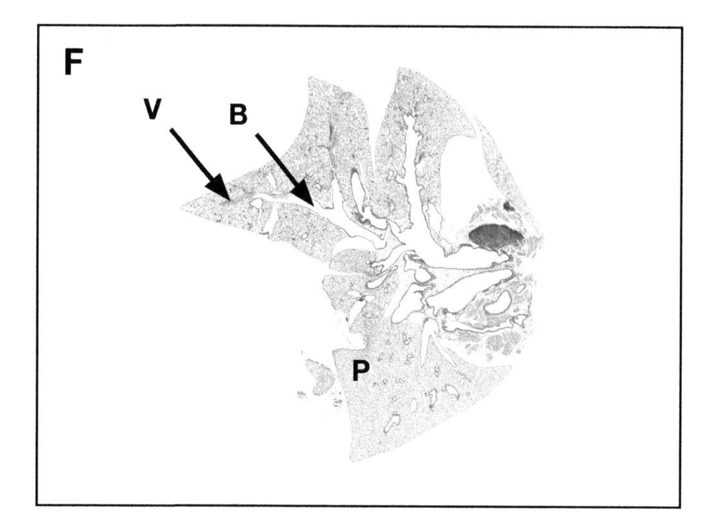

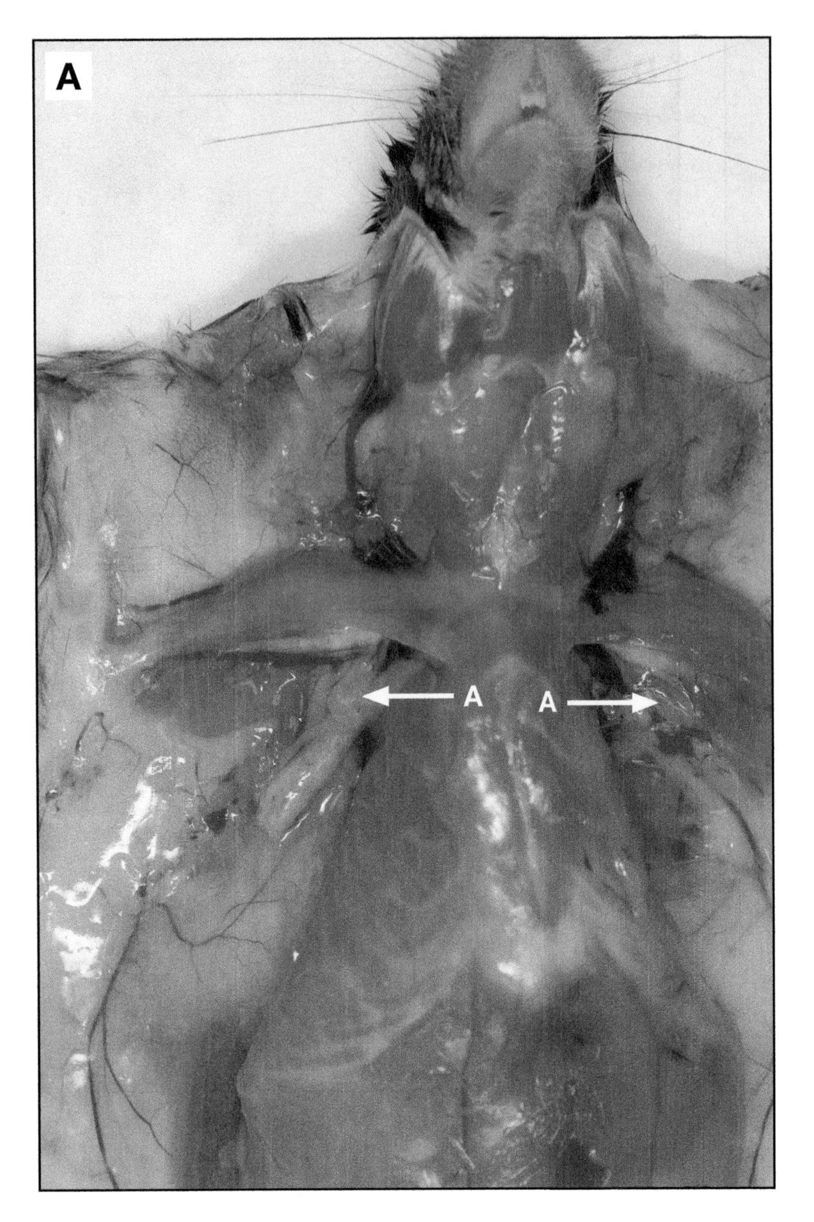

Lymph Nodes - Axillary

A. Axillary lymph nodes (A) are located in the armpits of the forelimbs. They are usually smoother and whiter than the pink grainy fat surrounding them.

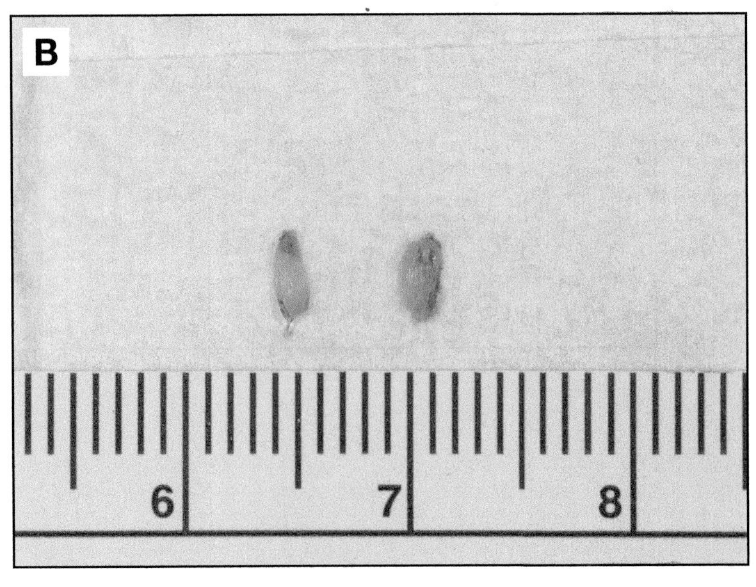

B. Axillary lymph nodes are small and should be placed into a CellSafe™ Biopsy Insert or wrapped in a piece of paper towel prior to cassetting.

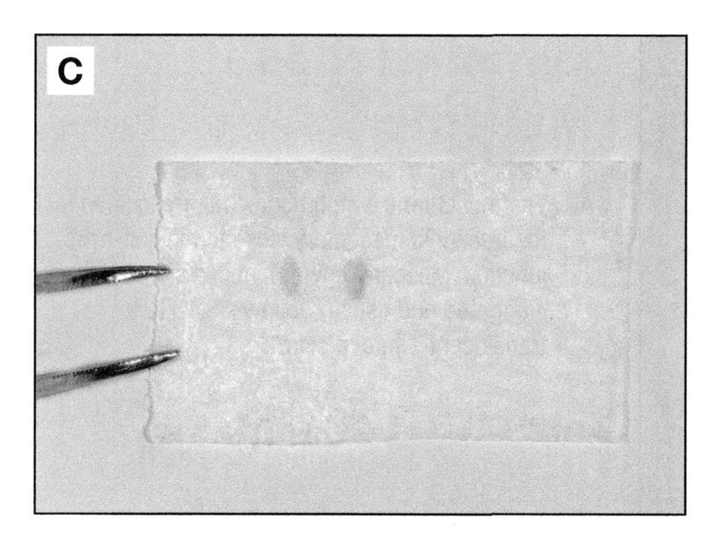

C. After processing, the lymph nodes will be slightly smaller and more translucent.

D. When embedding the lymph nodes, ensure that they are as flat as possible in the bottom of the mold. If there is fat around the nodes, try to first distinguish the node from the fat then place node face down.

E. Use care when facing the blocks as the axial lymph nodes are very small tissues.

F. The ideal section will contain lymphoid tissue from the axillary lymph nodes (A).

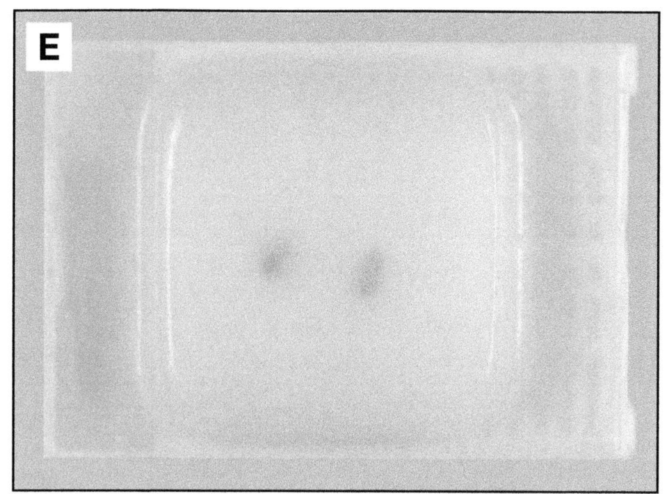

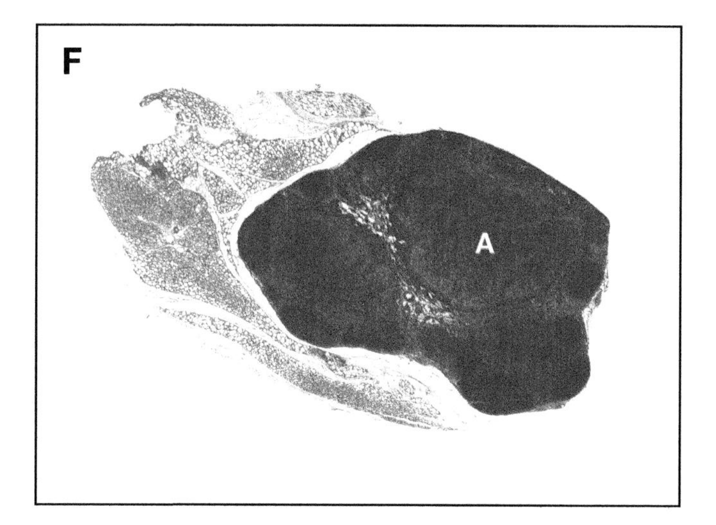

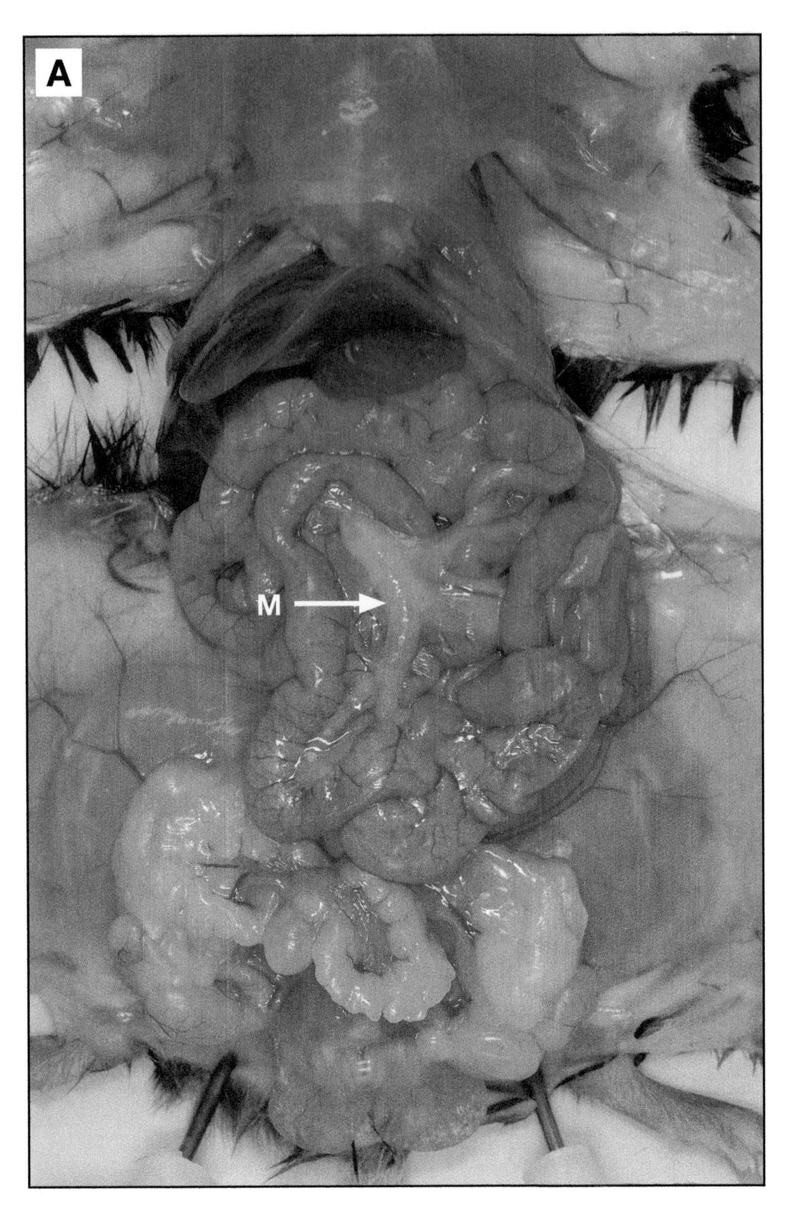

Lymph Nodes - Mesenteric

A. The mesenteric lymph nodes (M) are located in the intestinal mesentery close to the ileo-cecal junction. Mesenteric lymph nodes are smooth, elongated and usually yellowish-white to translucent-white in color.

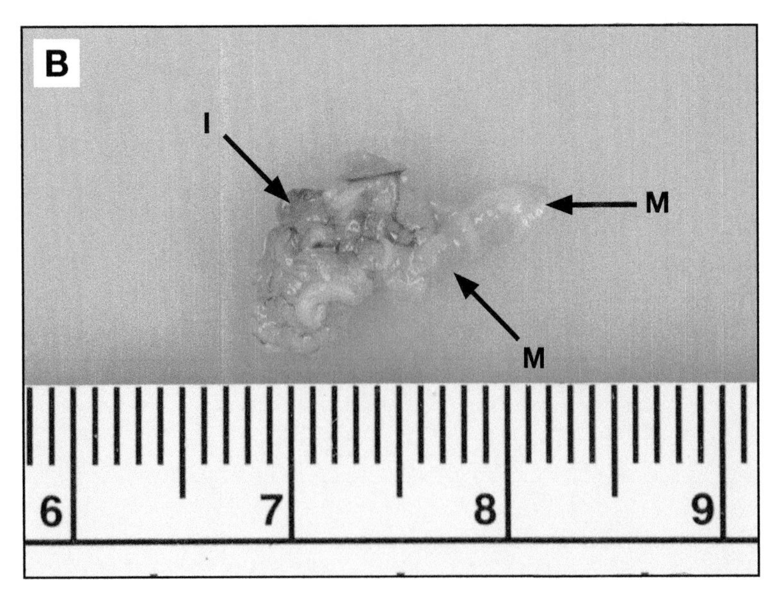

B. The mesenteric lymph nodes (M) have been removed with the intestinal mesenteric tissue. If the nodes are large, they can be removed and processed without the intestinal mesenteric tissue (I).

C. After processing, the mesenteric lymph nodes may be easier to see. For pathological analysis they do not need to be separated from the fat and mesentery.

D. When embedding the lymph nodes, ensure that they are as flat as possible in the bottom of the mold. Try to first distinguish the node from the fat and place the node down. If only lymph nodes (without fat) are present, gently press them to the bottom of the mold.

E. When facing into the block, it may be necessary to examine sections microscopically to see if the small dense nodes are present in the section. The mesenteric lymph nodes may be difficult to see within the fat surrounding them.

F. The ideal section will contain lymphoid tissue from the mesenteric lymph nodes (M). Pancreas (P) and intestinal mesenteric tissue (I) may also be present in these sections.

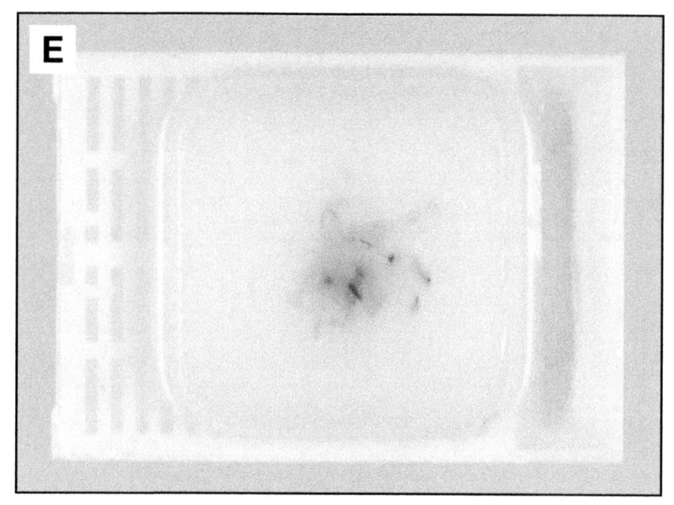

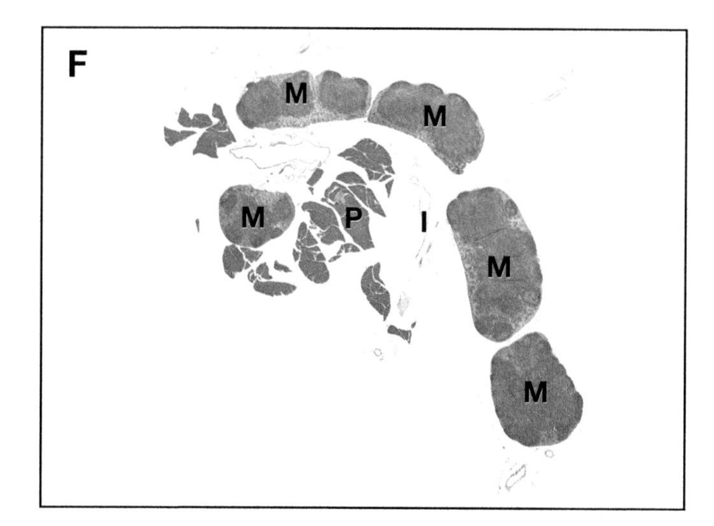

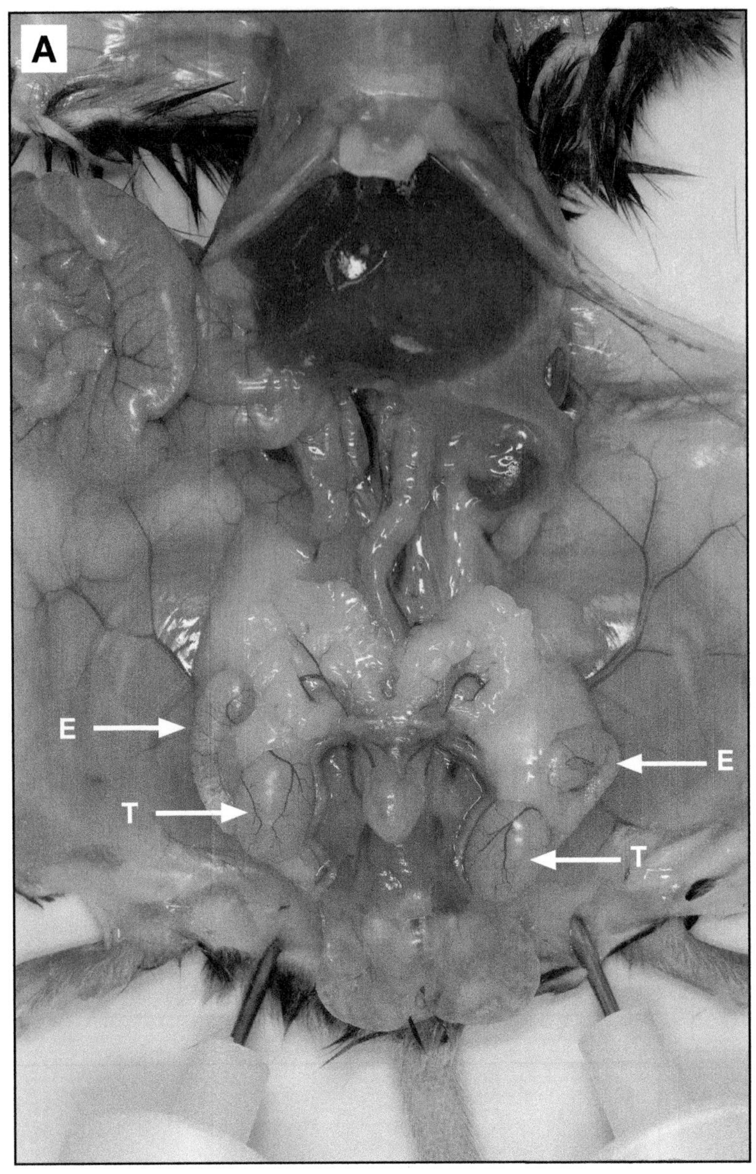

Male - Epididymis

A. The testes (T) are in the scrotal sac located in the lower abdomen. The epididymes (E) may need to be gently drawn out of the scrotal sac with the testes in order to be removed from the mouse.

B. The epididymes (E) are attached to the testes (T) and the organs on each side may be collected as a unit for histology.

C. Both epididymes are wrapped in a piece of paper towel to hold them straight and flat during fixation and processing.

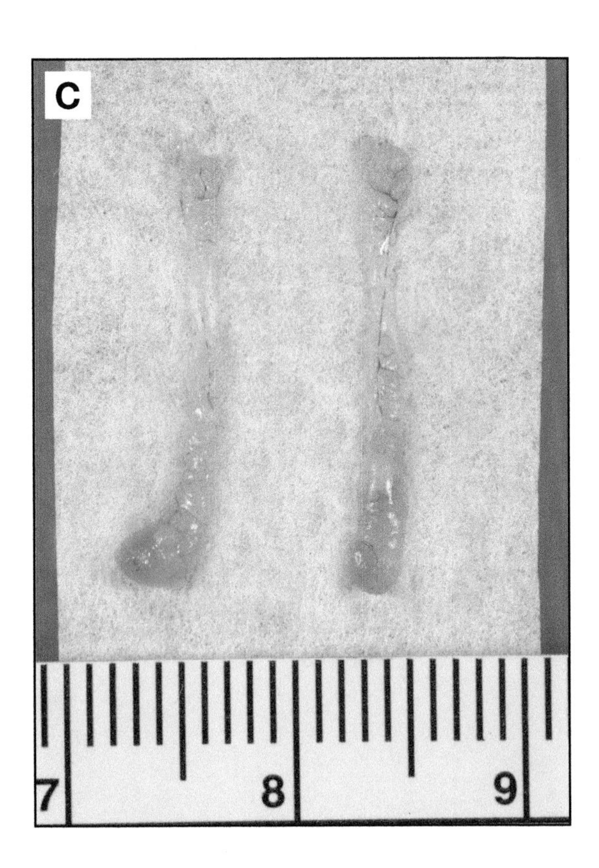

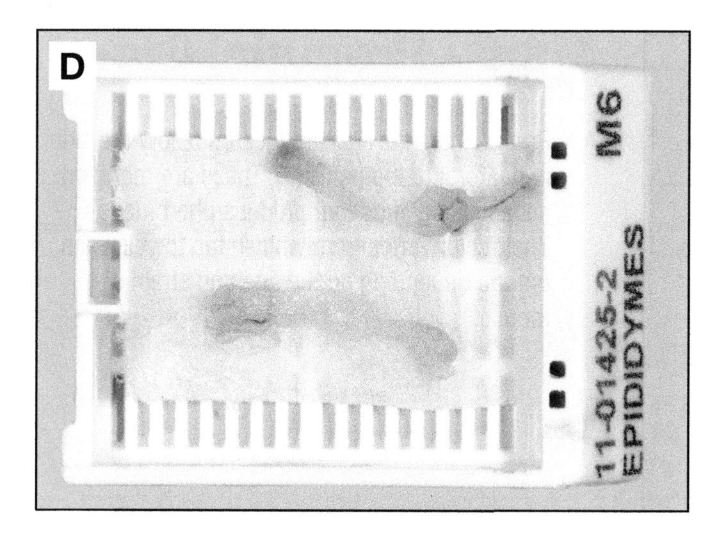

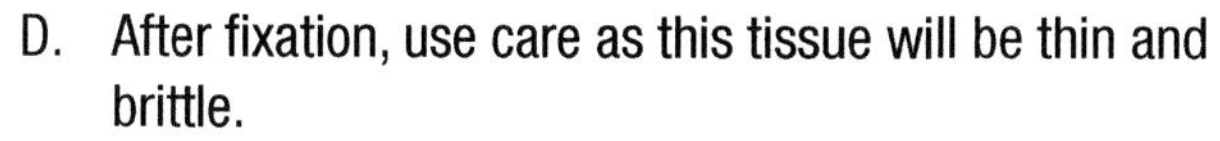

D. After fixation, use care as this tissue will be thin and brittle.

E. The epididymes should be embedded with the long face down, as flat to the bottom of the mold as possible.

F. Trim into the block until a full face from each epididymis can be seen. Because they are thin, it may not be possible to get the whole length of both epididymes in one section.

G. Sections for analysis should contain as much of a full face as possible of each epididymis.

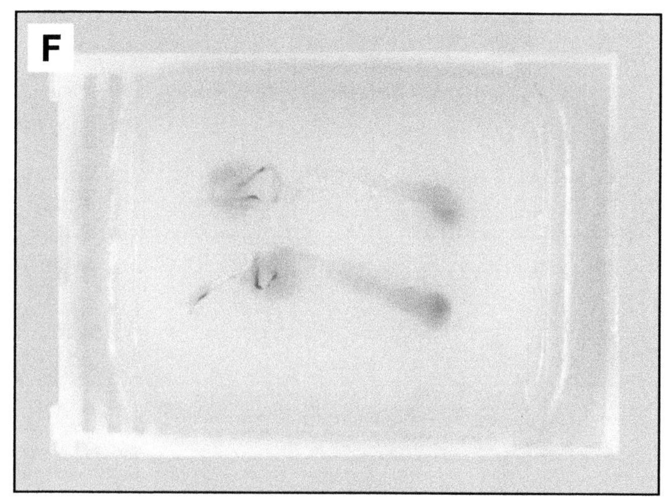

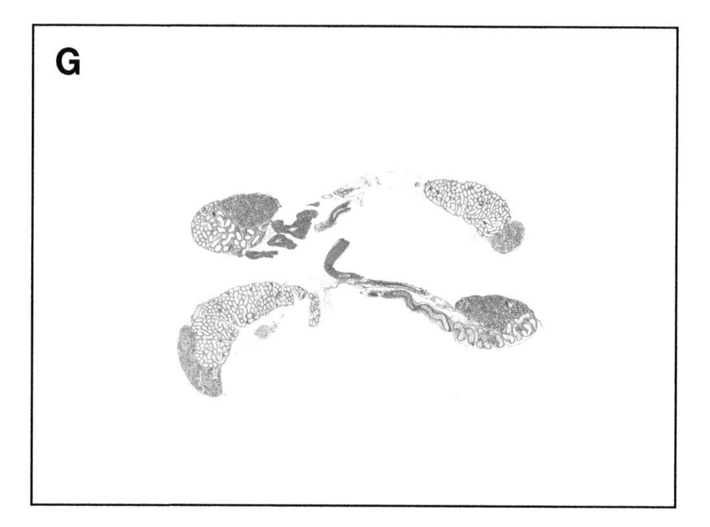

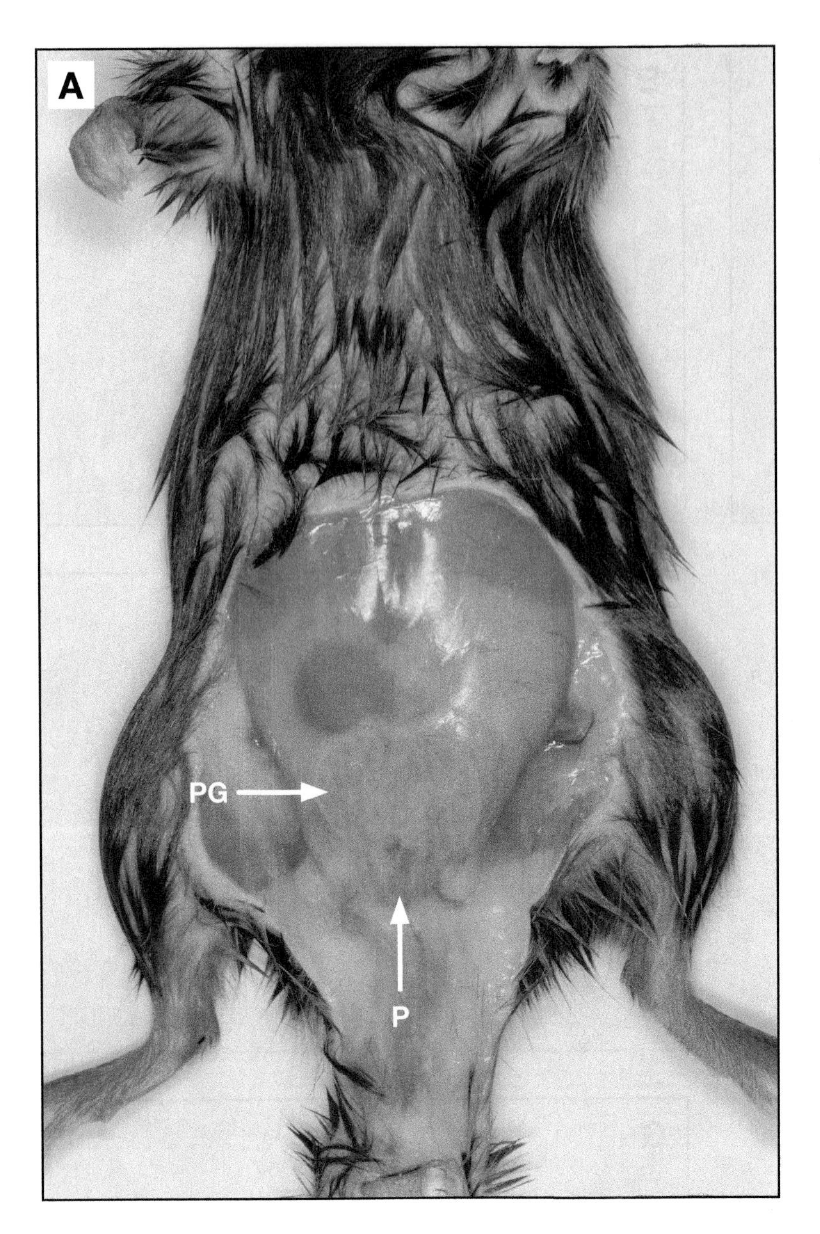

Male - Preputial Gland

A. Preputial glands (PG) are located below the skin and lateral to the penis (P). These are modified sebaceous glands that produce pheromones. Their color varies from whitish tan to yellowish orange depending on the age and strain of mouse.

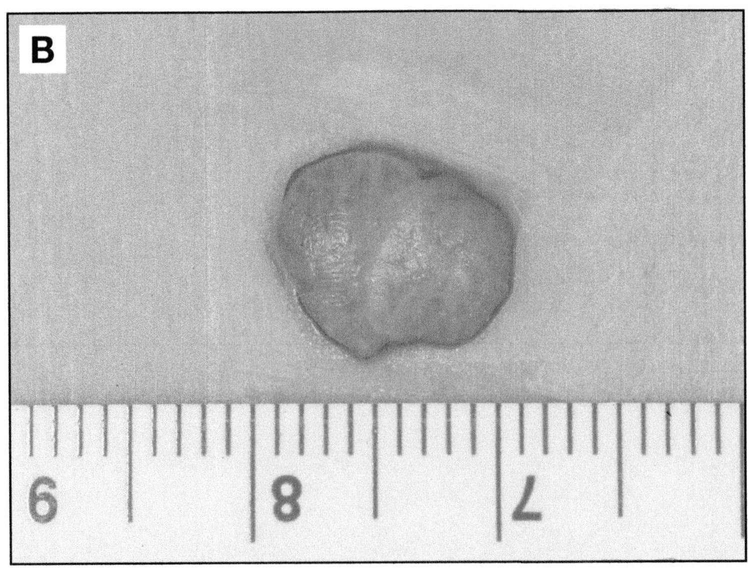

B. Preputial glands are paired organs and should be removed together. Be sure they are flat when cassetting in order to ensure proper embedding later.

C. After processing, the preputial glands are smaller and darker in color.

D. Since the preputial glands are homogeneous, they may be placed flat into the embedding mold with either the ventral or dorsal side down.

E. Face into the block until a full face of the entire gland is visible.

F. A good section will contain large excretory ducts (D) surrounded by acini (A) composed of eosinophilic secretory sebaceous cells and elongated basal cells.

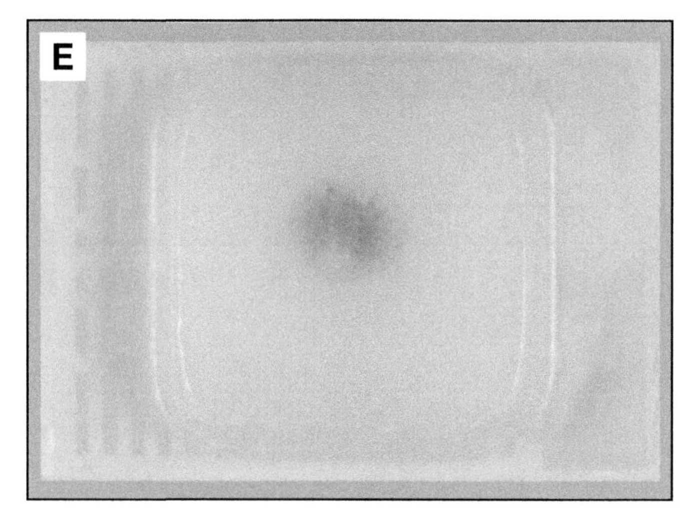

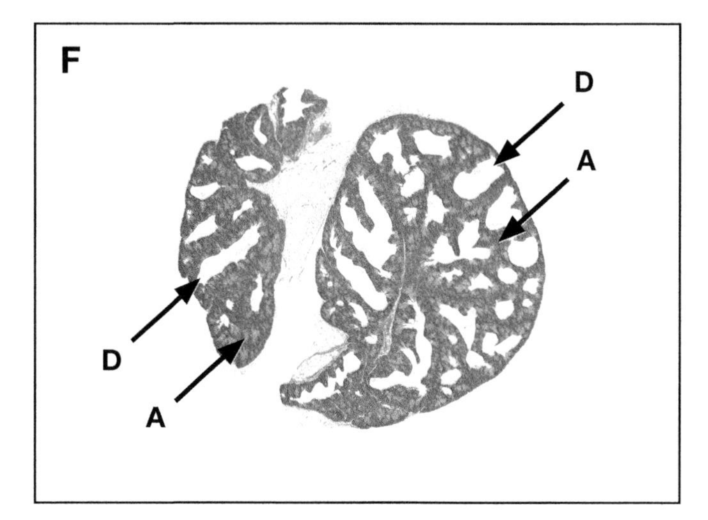

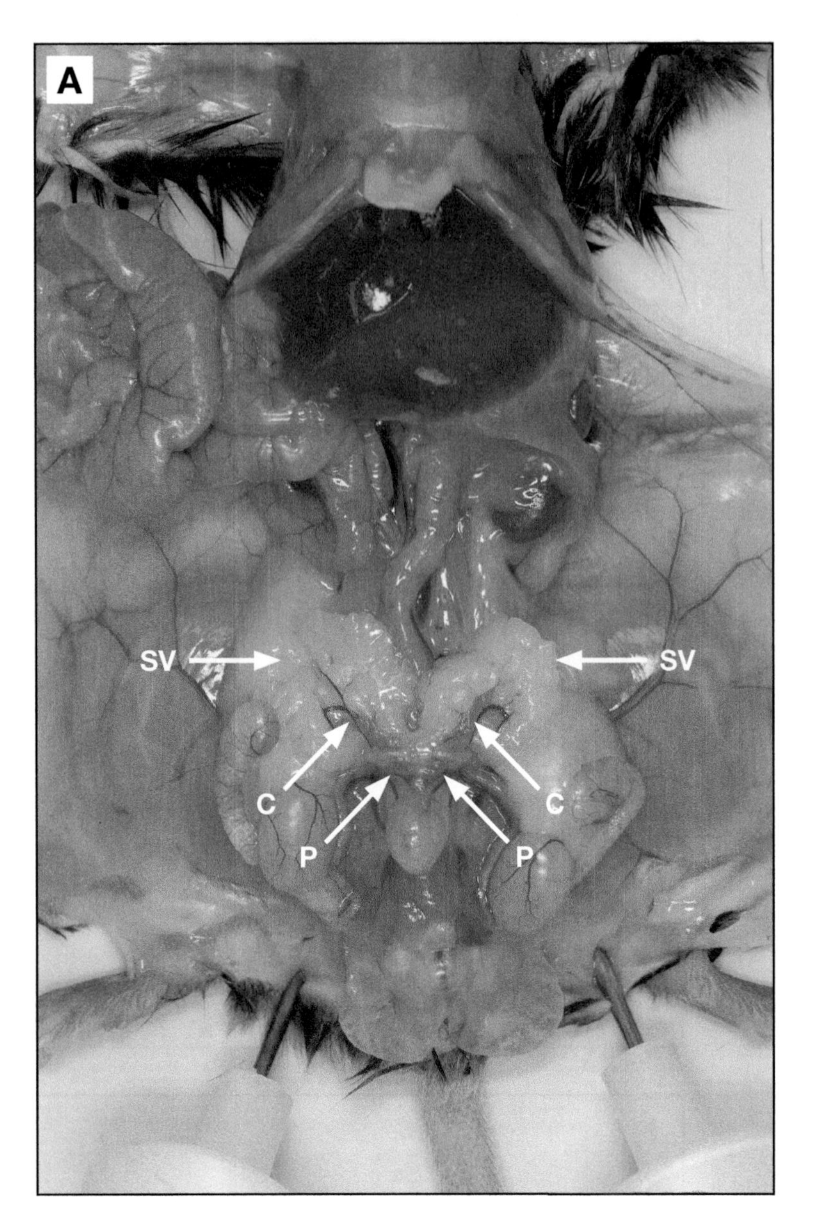

Male - Seminal Vesicle, Coagulating Gland, Prostate Gland

A. The seminal vesicle (SV) is a white ram's horn shaped organ in the lower end of the abdominal cavity and usually located below the intestines. The prostate gland (P) sits on top and below where these horns come together. The coagulating glands (C) are found on the inside of the seminal vesicles.

B. Remove the seminal vesicles (SV), prostate (P) and coagulating glands (C) together as a unit. Lay this unit on a piece of paper towel to maintain organ orientation during fixation and processing. If possible, keep the tiny, white tubes of the ureters (U) attached because they can be used to distinguish the ventral sides [of tissues].

C. Trimming can be done at any time prior to embedding. Here, before fixation, two transverse sections are made through the widest part of seminal vesicle and coagulating glands on each side. The resulting seminal vesicle pieces (arrows) and central piece (*) containing prostate gland are processed together.

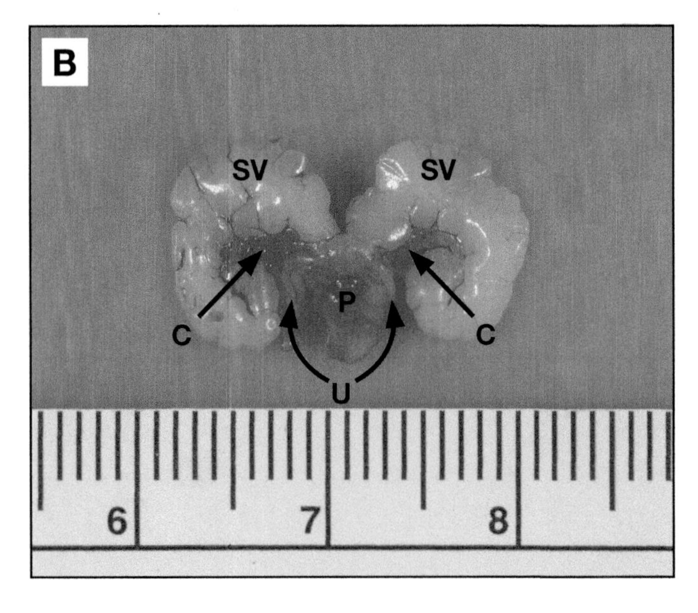

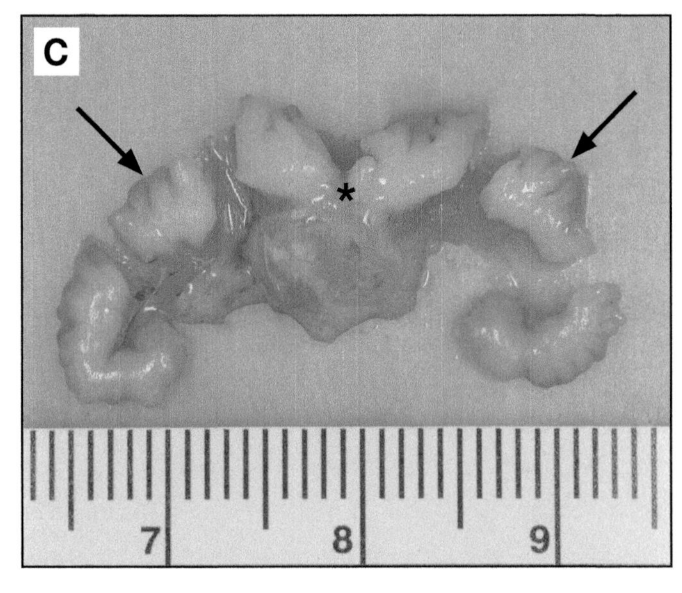

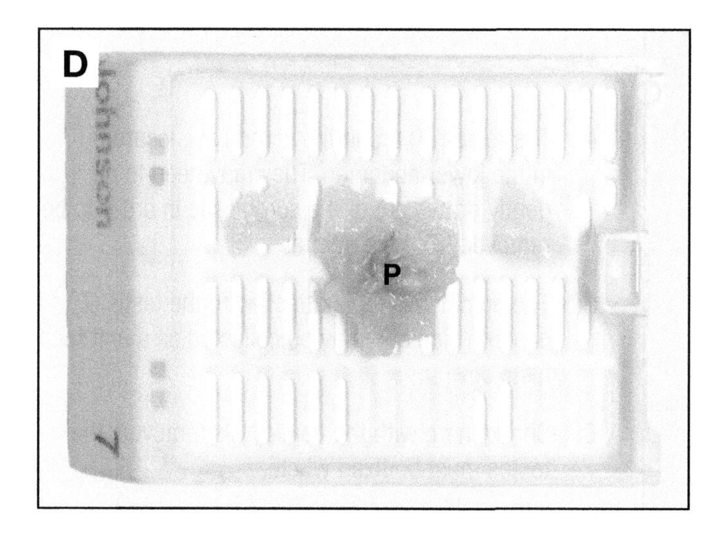

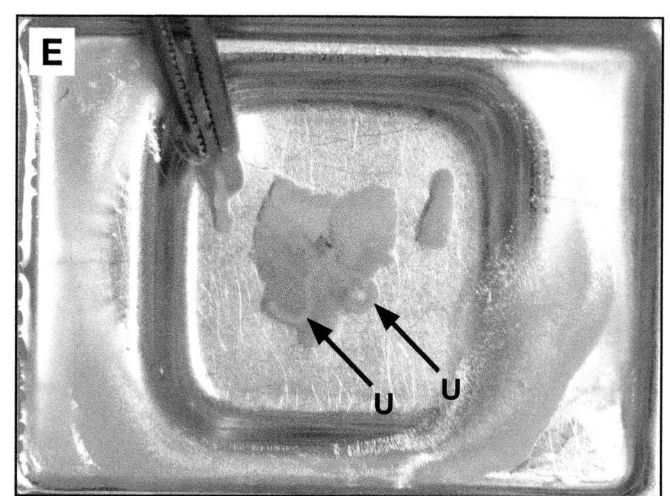

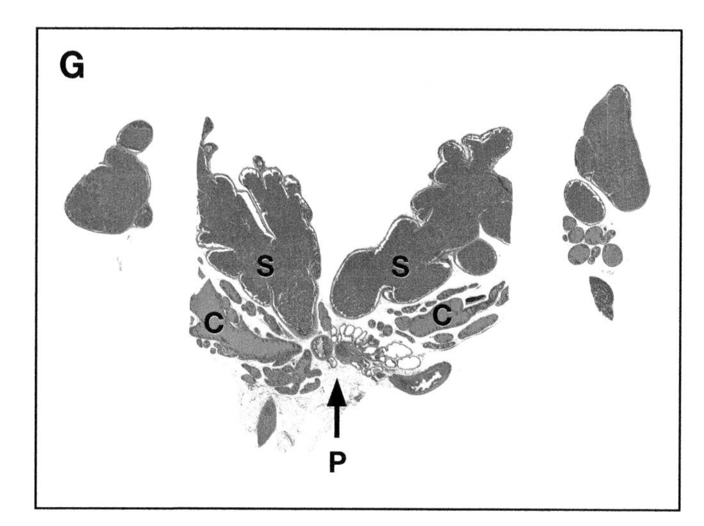

D. After processing, the prostate gland (P) becomes a darker color and easier to distinguish from the seminal vesicles.

E. Embed the large piece containing the prostate gland ventral side and ureters (U) facing up, flat into the mold. The two separate pieces of the seminal vesicles and coagulating gland are embedded cut face down.

F. Face into the block to obtain a full view of all three embedded pieces. It may be necessary to examine sections microscopically to ensure that prostate gland is present.

G. A good section must contain seminal vesicle (S), coagulating gland (C), and prostate gland (P).

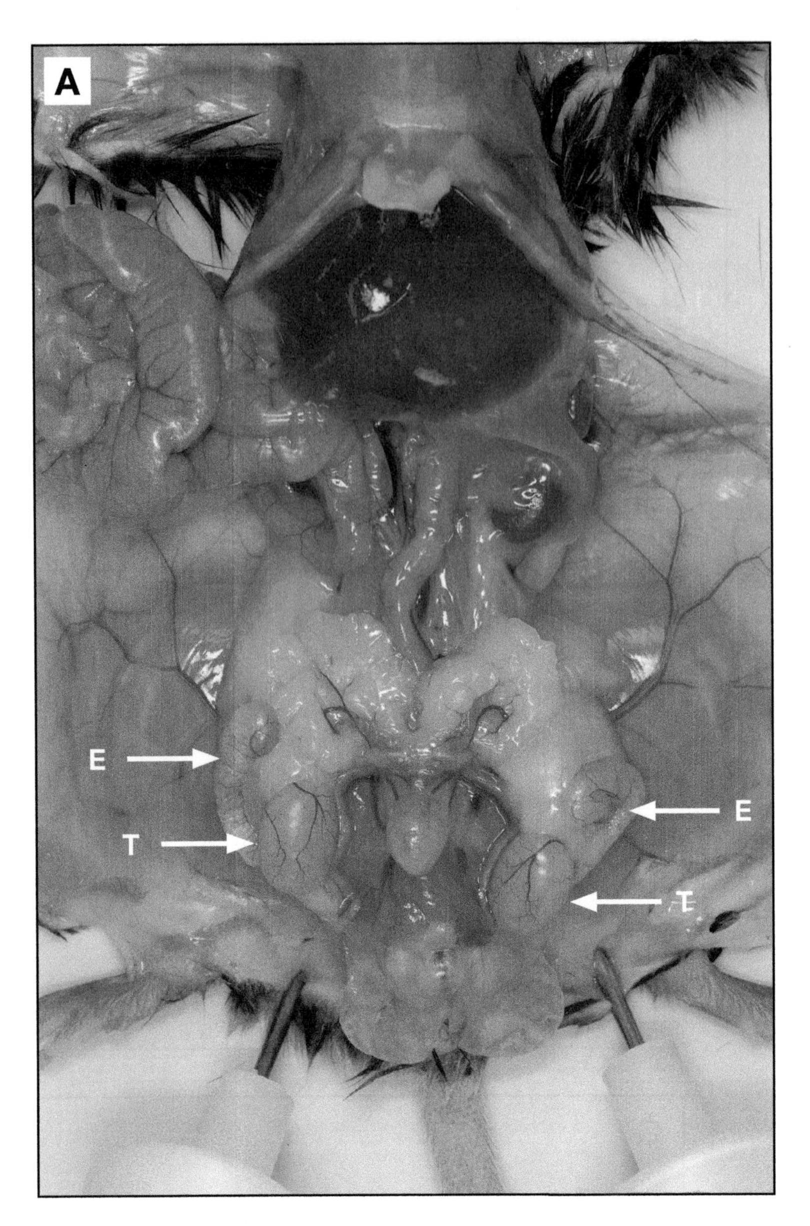

Male - Testes

A. The testes (T) are in the scrotal sac located in the lower abdomen. They may need to be gently drawn out of the scrotal sac in order to be removed from the animal.

B. The epididymis (E) is attached to the testis (T) on each side and may be collected as a unit for histology.

C. Shown here with the epididymis removed, the testes must be fixed whole.

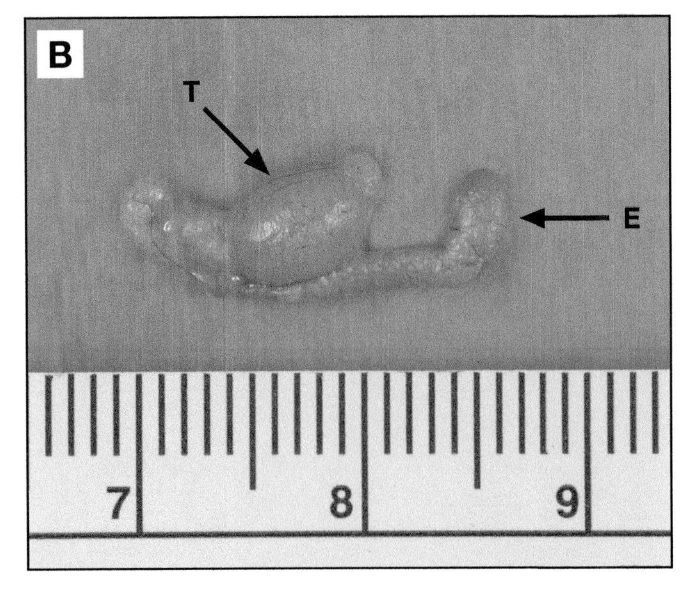

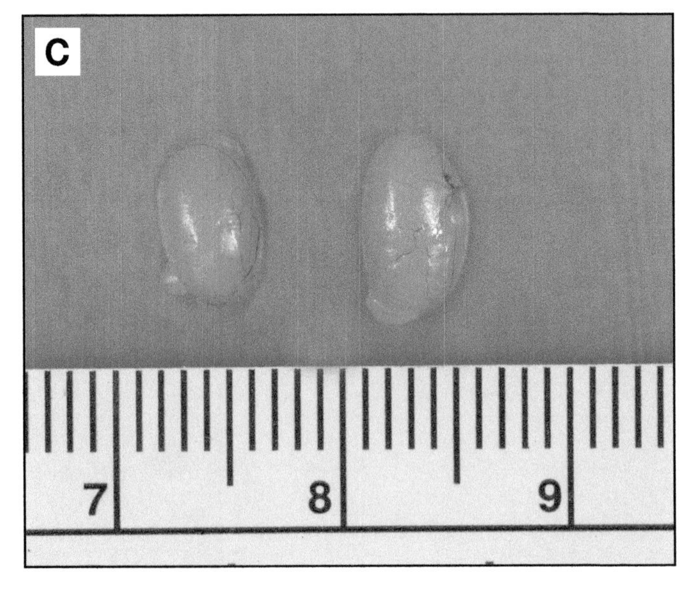

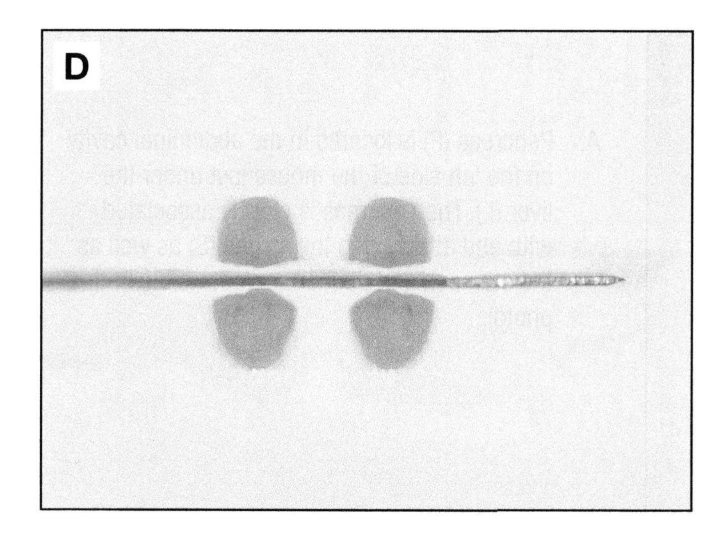

D. The testes can be trimmed either after fixation or after processing as shown here. The testes are cut in cross section.

E. A minimum of one half of each testis is embedded with cut faces down in the mold.

F. Face into the block until a full face of each testis piece is seen.

G. Sections for analysis should contain a full cross section of each embedded testis.

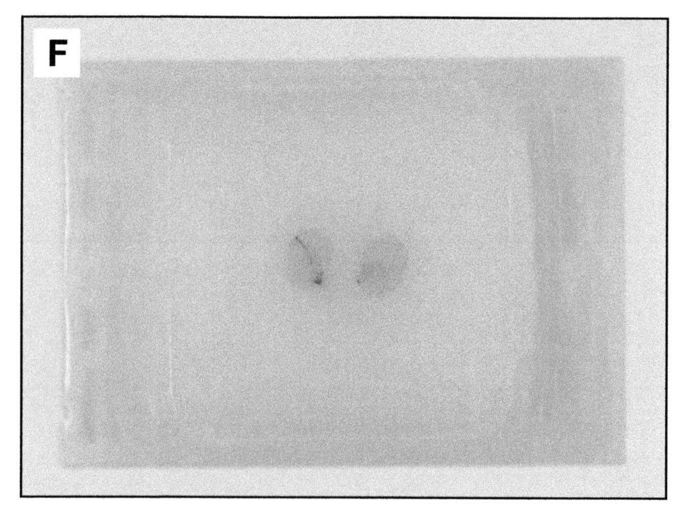

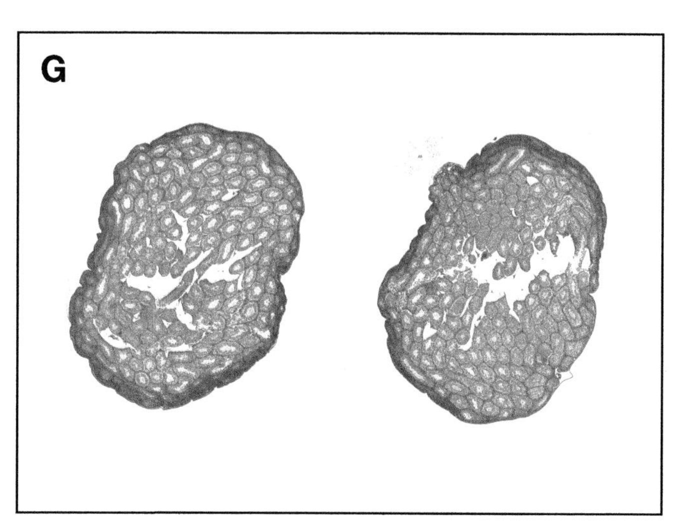

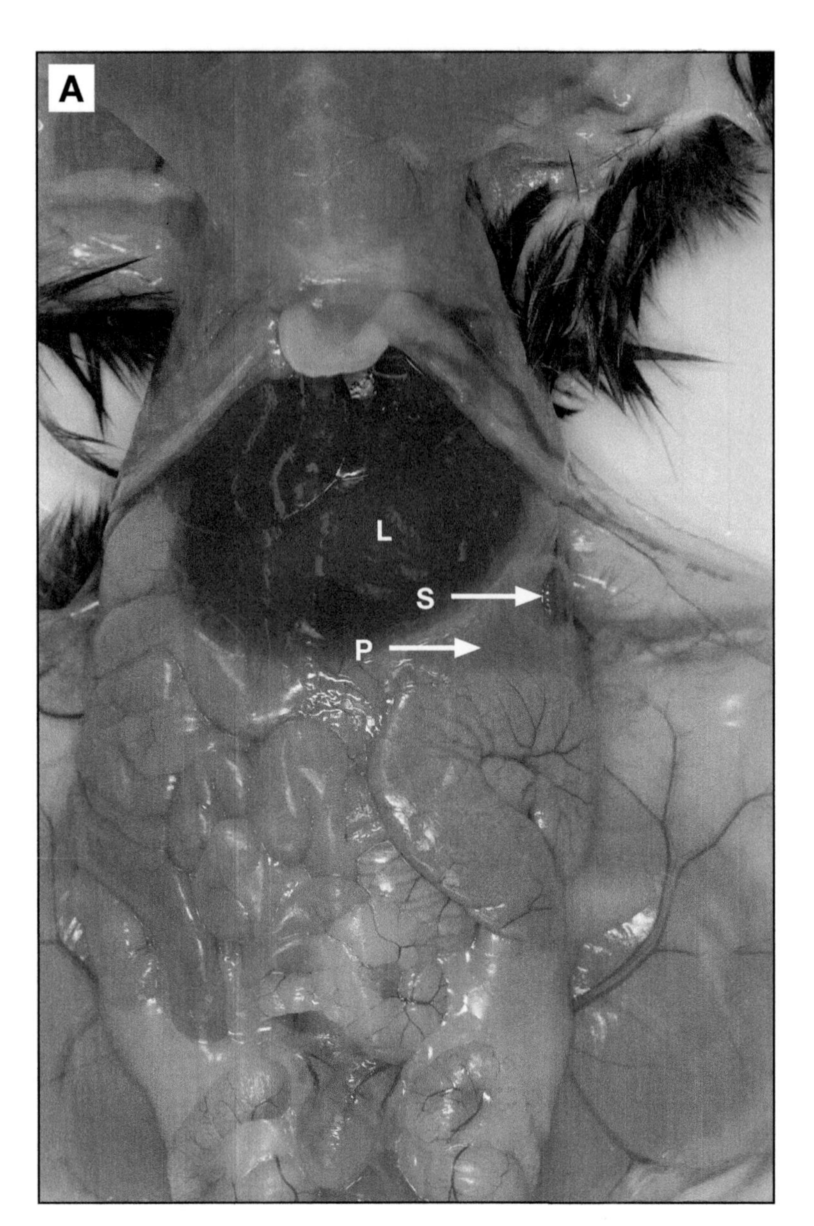

Pancreas

A. Pancreas (P) is located in the abdominal cavity on the left side of the mouse just under the liver (L). The pancreas is closely associated with and attached to the spleen (S) as well as the stomach and duodenum (not visible in this photo).

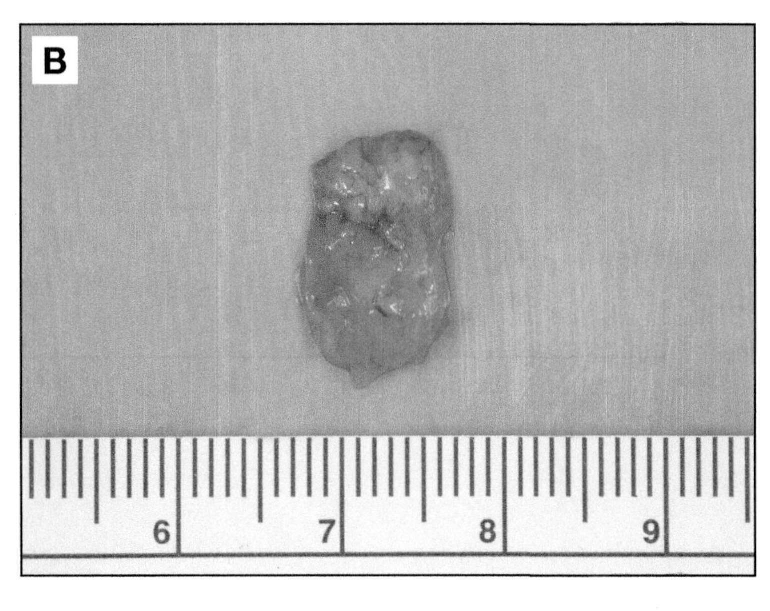

B. Upon removal, the pancreas can be cassetted without further trimming.

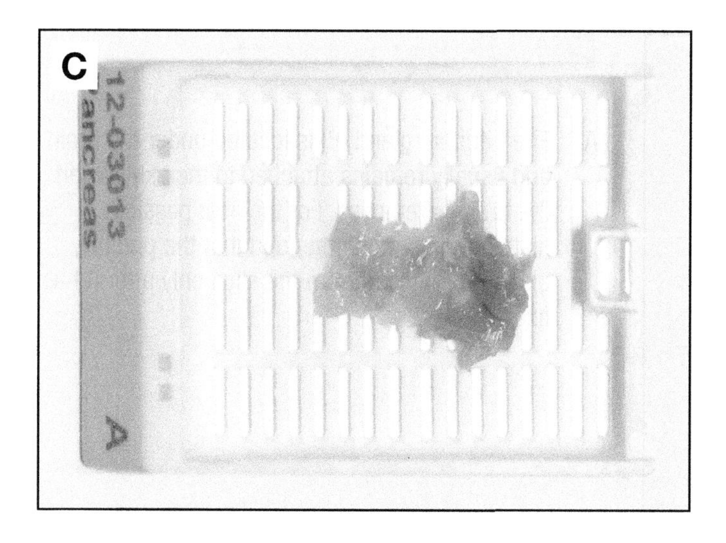

C. After processing, the pancreas will be smaller and darker in color.

D. Embed the pancreas so that maximum surface area is exposed face down into the mold.

E. The downward face of the block should reveal as much surface area of the tissue as possible.

F. A section of pancreas will look loosely connected to stromal tissue in the duodenal mesentery. Hematoxylin and eosin stained islet cells (islets of Langerhans (I)) are light pink in color with dark purple nuclei. Acinar cells (A) are dark purple with large zymogen granules. Lymph nodes (L) can sometimes be seen as well.

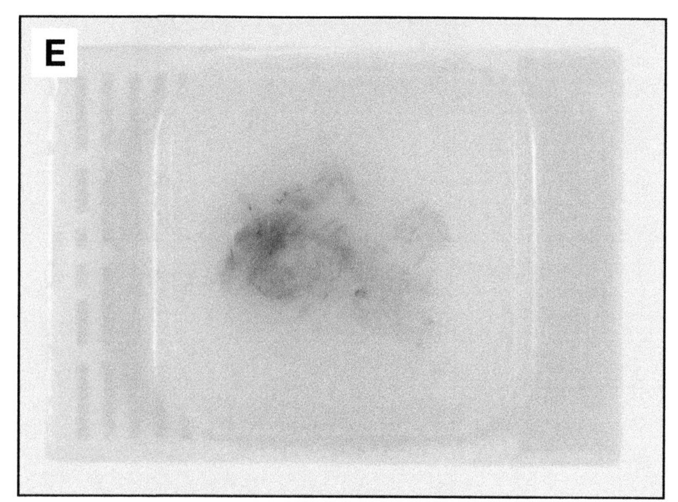

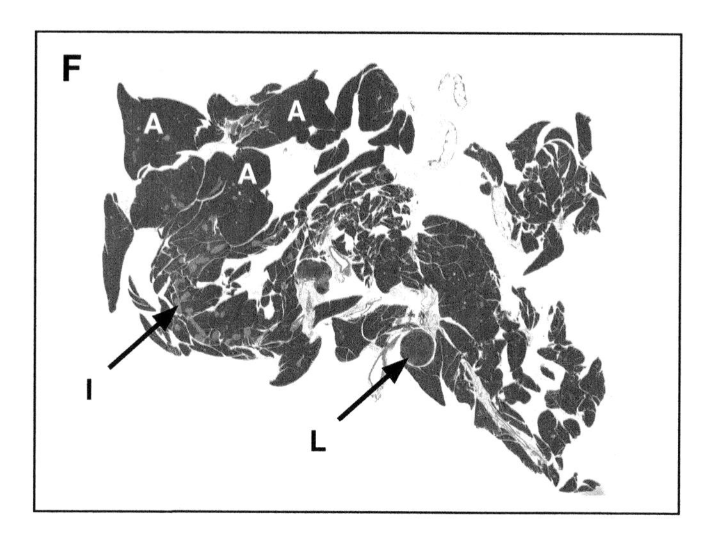

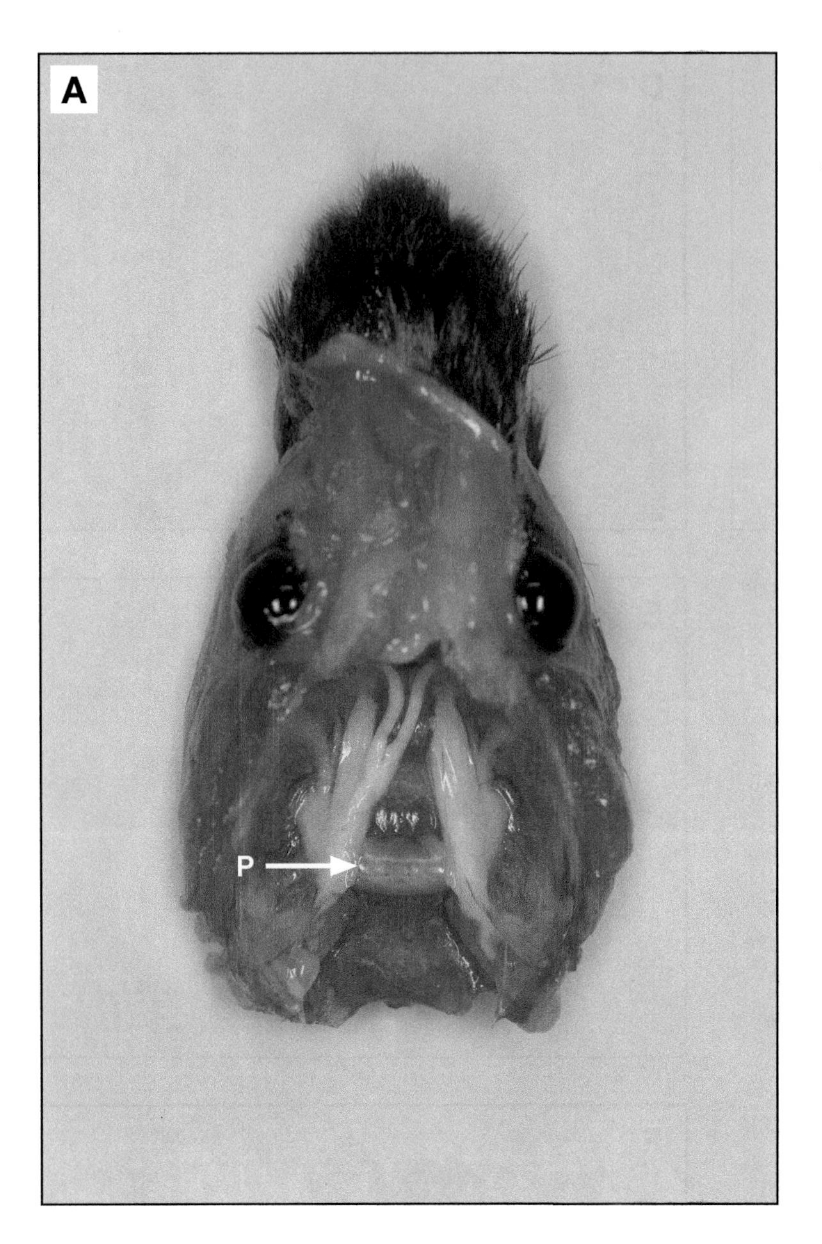

Pituitary Gland

A. The pituitary gland (P) is located under the brain and usually remains attached to the skull when the brain is removed. For the best possible histology, it is recommended that the pituitary gland be removed from the skull only after it has fixed in situ.

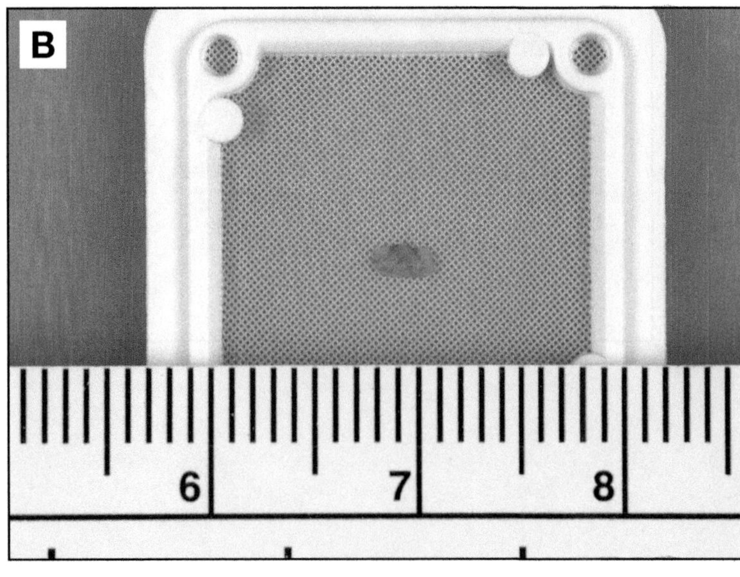

B. The fixed pituitary should be placed into a CellSafe™ Biopsy Insert or wrapped in a piece of paper towel prior to processing.

C. After processing, the pituitary gland is even smaller.

D. Care should be taken when embedding the pituitary gland to not separate the pieces of the intermediate and posterior lobes from the larger anterior lobe.

E. If possible, the pituitary should be embedded so a cross section is taken through the lobes of the gland.

F. The ideal section will show all three lobes – the anterior (A), intermediate (I) and neurohypophysis (N).

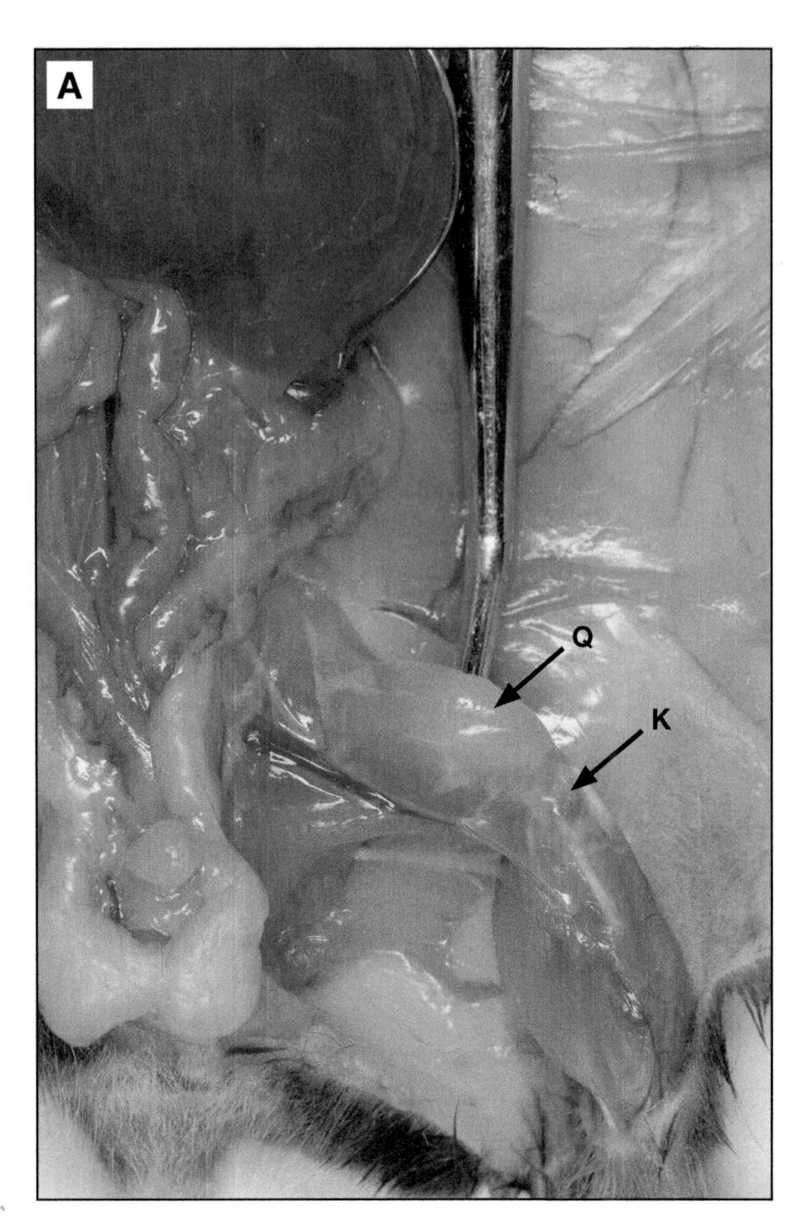

Quadriceps Muscle

A. The quadriceps muscle (Q) is located just under the skin on the hind leg above the femur. The quadriceps muscle originates at the hipbone (below muscle and not seen here) and inserts at the knee (K).

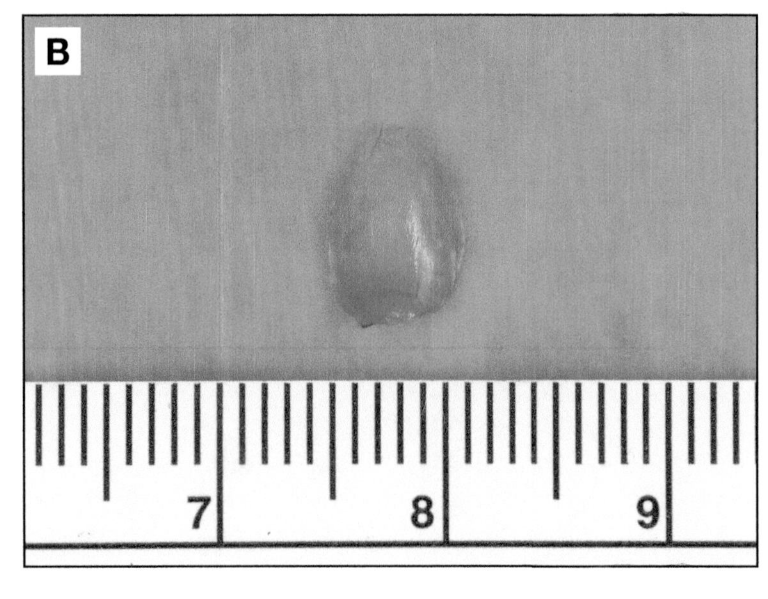

B. Try to retain the pointed end of the muscle where it attaches to the knee for easier orientation during embedding.

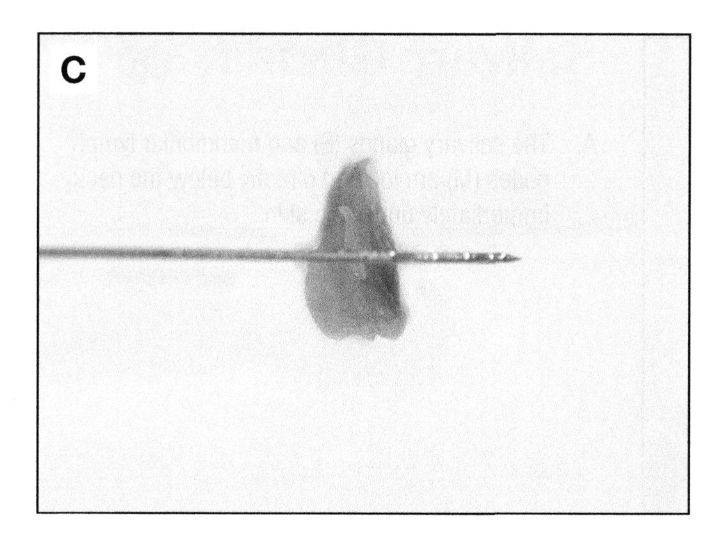

C. After processing, the muscle appears shrunken and darker in color. It should be cut in half across the short edge of the muscle.

D. To embed the quadriceps muscle, lay the pointed end piece on its side to obtain a longitudinal section of muscle fibers. The remaining piece is embedded cut face down for cross sections of the muscle bundles.

E. Face into the block far enough to obtain full cross and longitudinal surfaces of the muscle pieces.

F. An ideal section will demonstrate both cross (C) and longitudinal (L) sections through the quadriceps muscles.

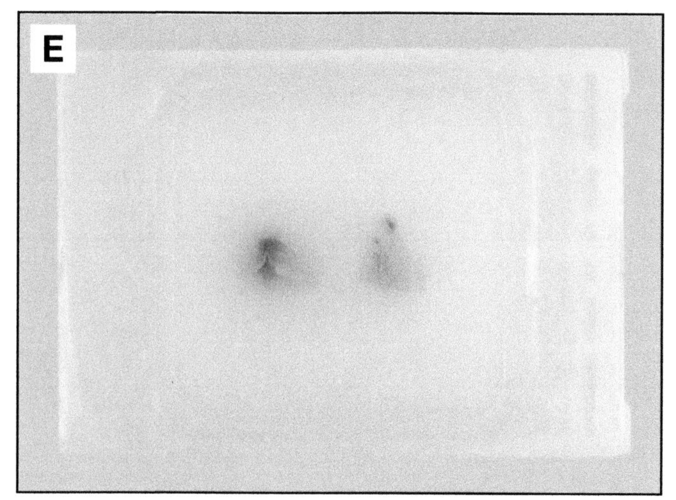

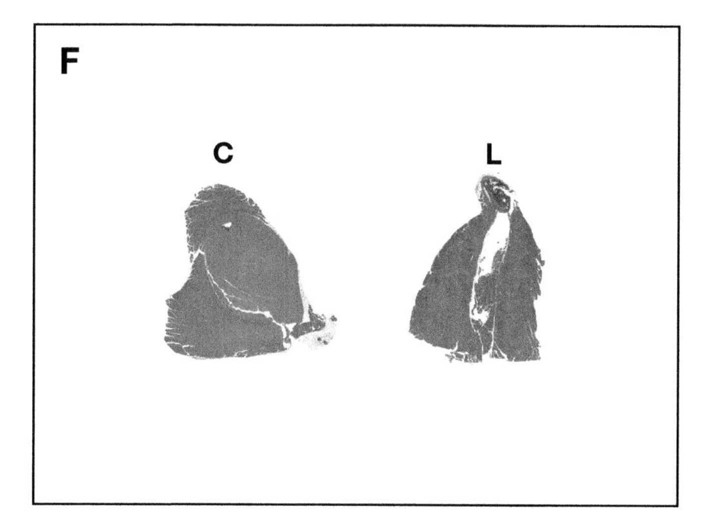

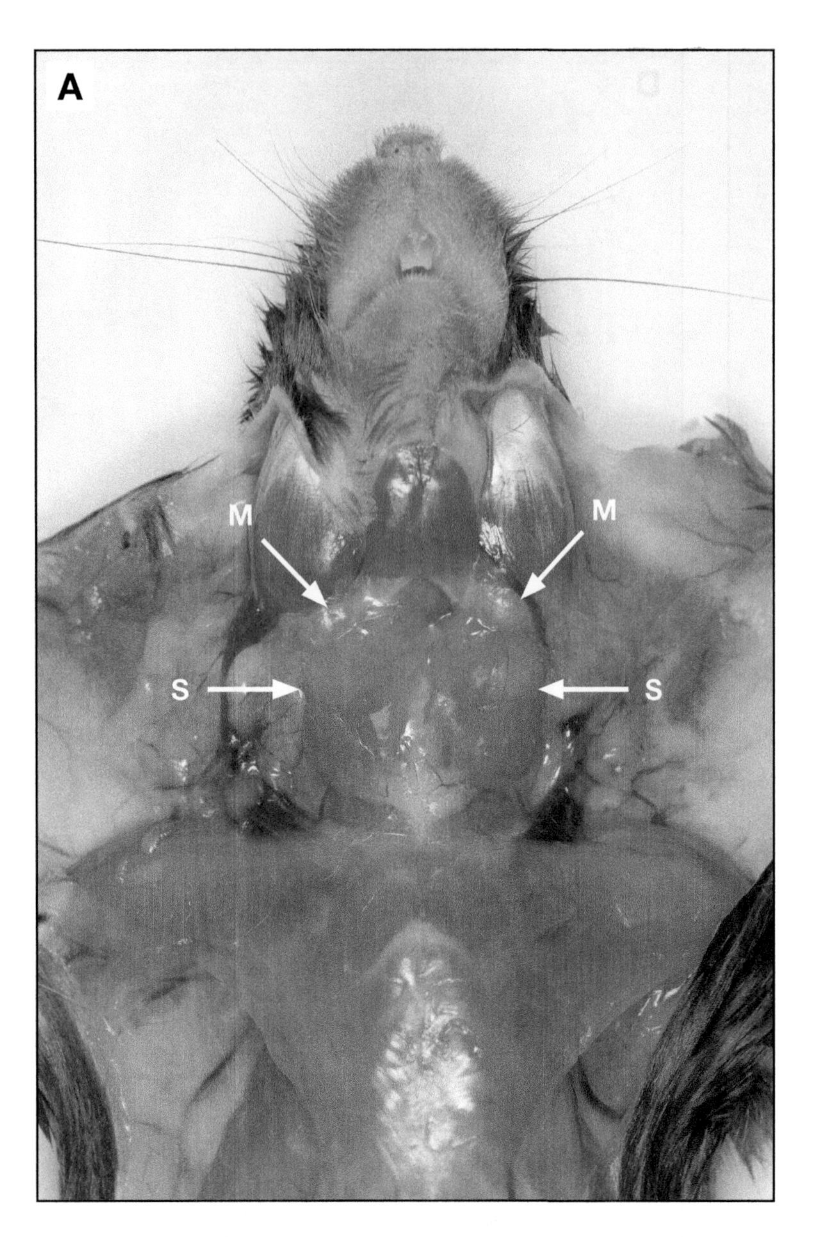

Salivary Glands

A. The salivary glands (S) and mandibular lymph nodes (M) are located directly below the neck immediately under the skin.

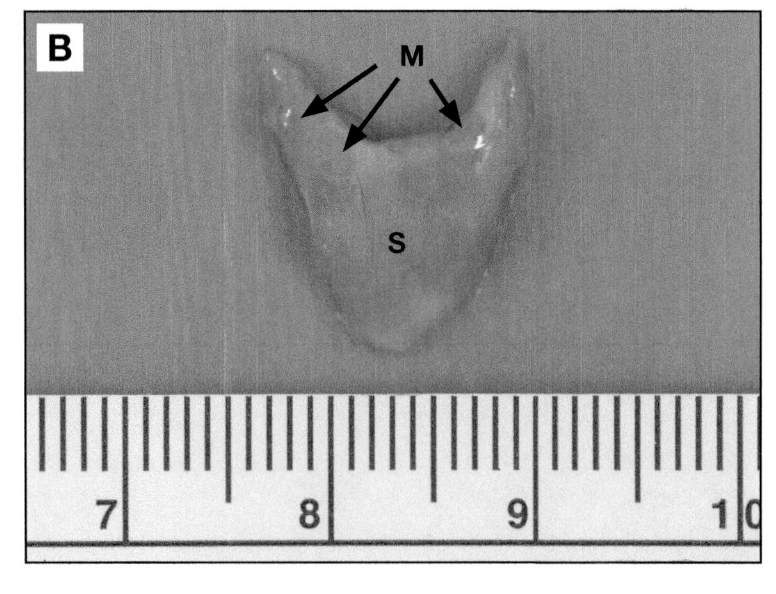

B. It is best to remove the salivary glands (S) and the mandibular lymph nodes (M) together as a unit, shown ventral side up in this photo.

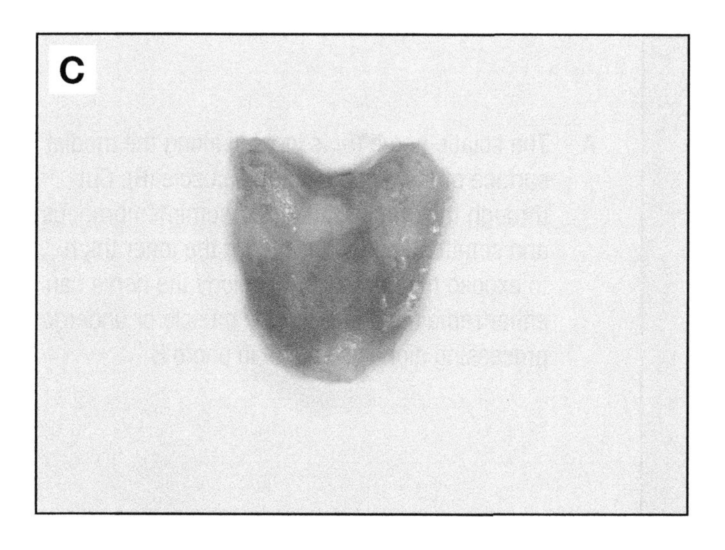

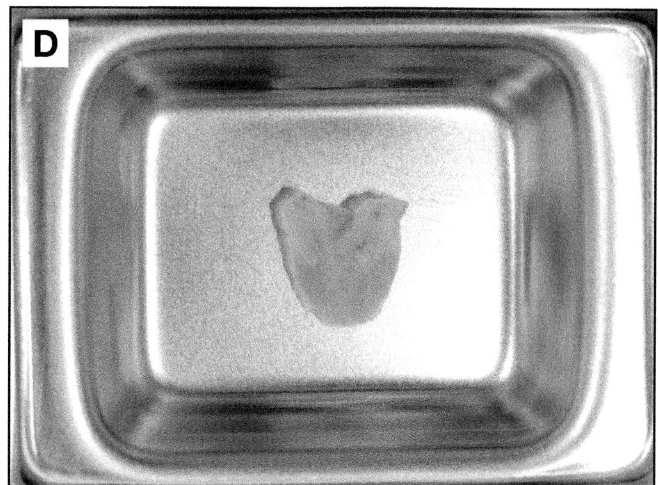

C. After processing, the glands appear darker and the lymph nodes appear whiter.

D. Place the salivary glands and mandibular lymph nodes ventral side down into the mold at embedding.

E. When facing into the block, you may need to view sections microscopically to ensure that the lymph nodes are present.

F. If embedded together as a unit, a section must contain at least one lymph node (L) and an area of salivary gland (S).

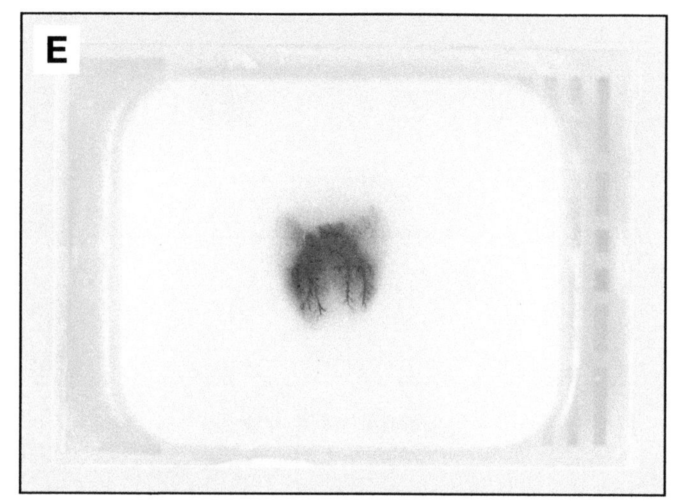

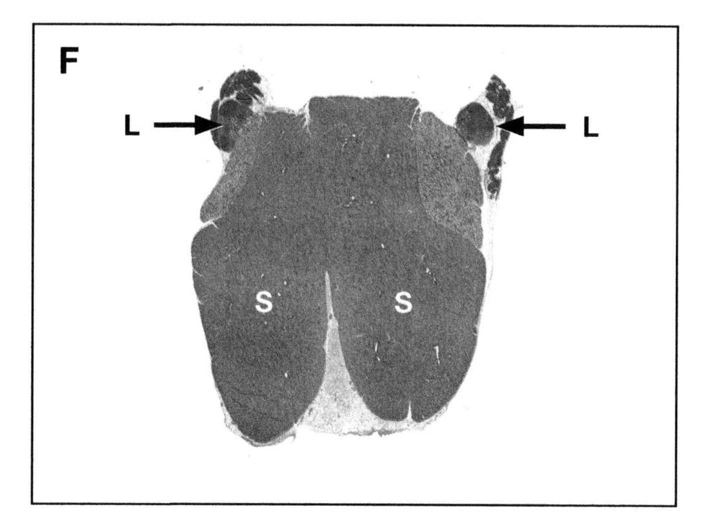

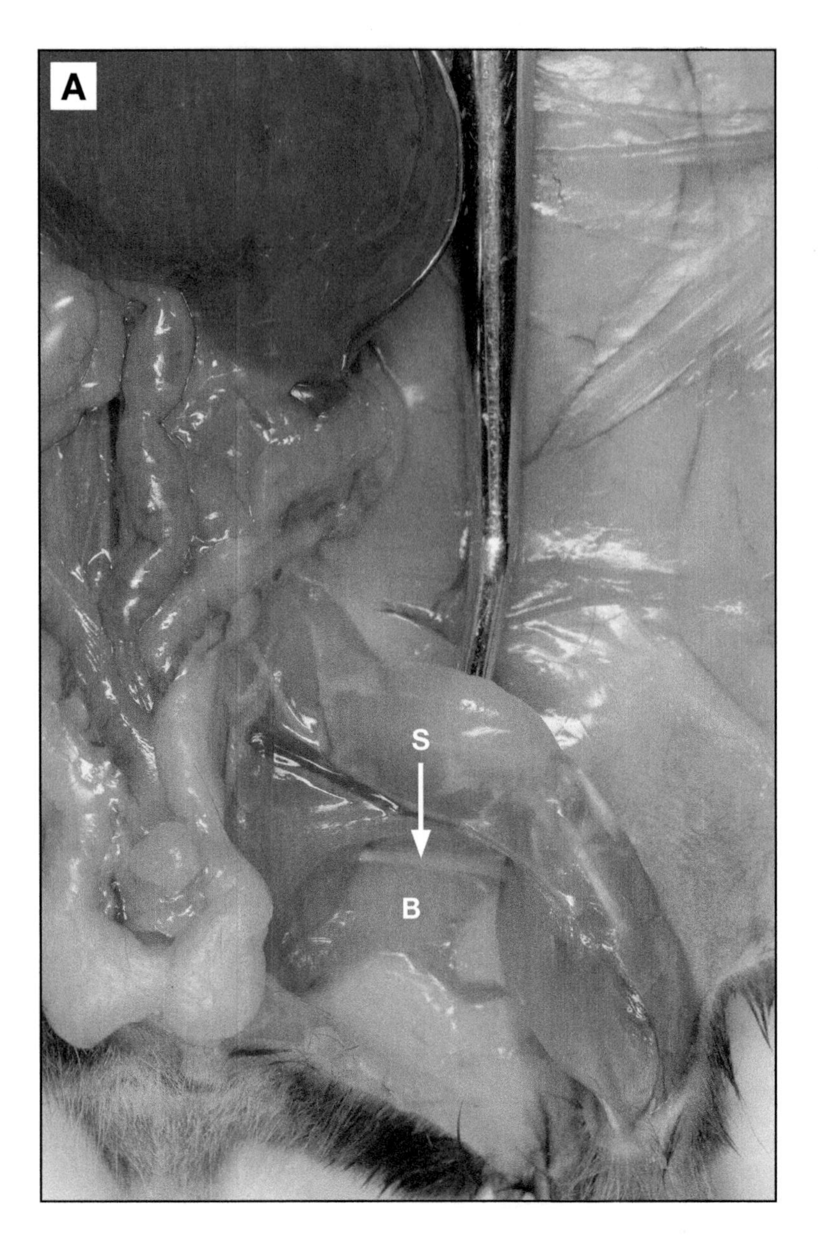

Sciatic Nerve

A. The sciatic nerve (S) is located along the medial surface of the biceps femoris muscle (B). Cut through the gracilis, adductor, semimembranous and semitendinous muscles on the inner thigh to expose the nerve. For histology the nerve can either remain attached to the muscle or undergo processing alone as shown in photo B.

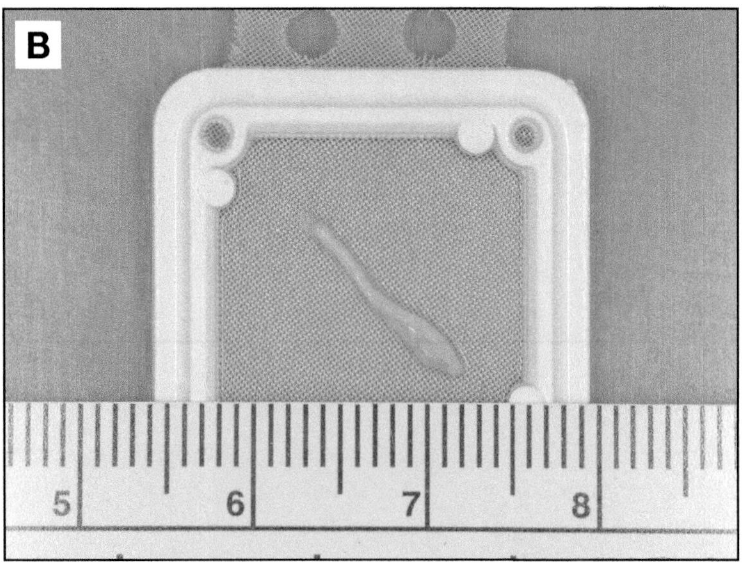

B. Place the delicate sciatic nerve into a CellSafe™ Biopsy Insert or wrap in a piece of paper towel to keep the small nerve straight during fixation and processing. This will make orientation at embedding easier.

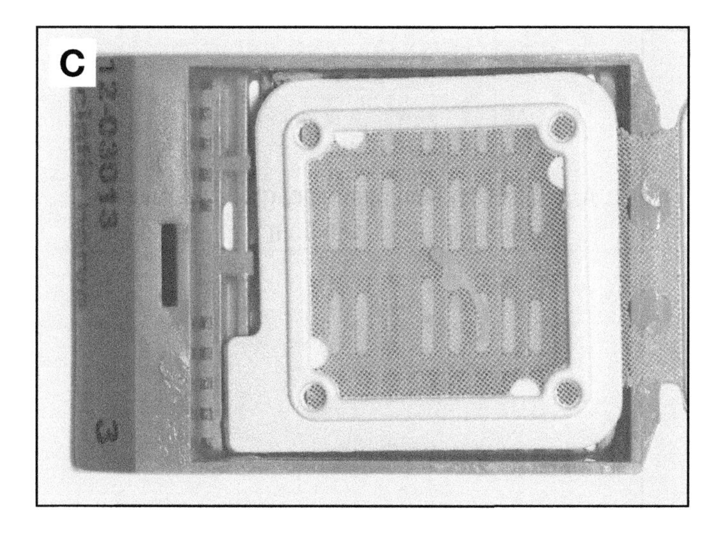

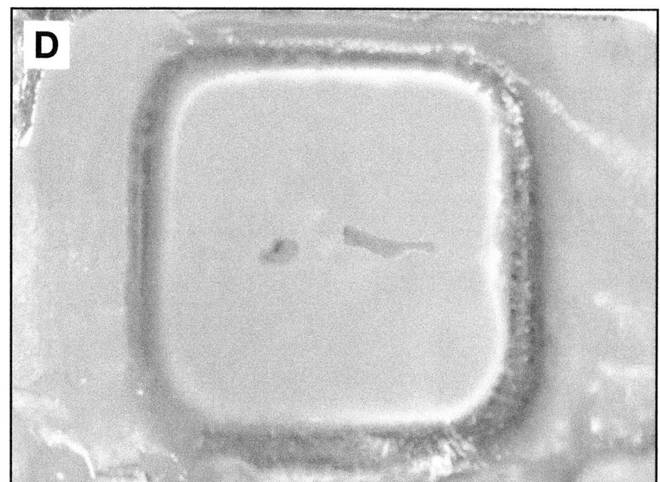

C. After processing cut the nerve into two pieces, roughly 1/3 and 2/3 lengths.

D. At embedding, lay the longer piece flat on its side in the base of the mold to show a longitudinal section of the nerve fibers. Stand the short piece on its cut face to obtain a cross section.

E. Facing into the block must be done carefully to obtain both a longitudinal and cross section without cutting through the thin nerve.

F. The ideal section contains both a longitudinal (L) and cross (C) section of the nerve.

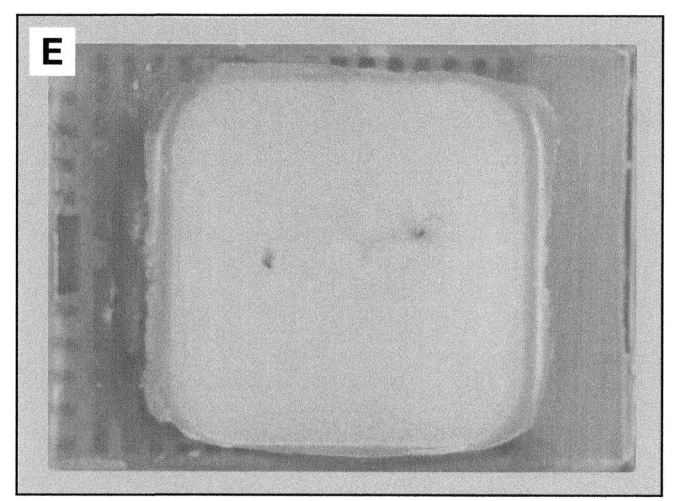

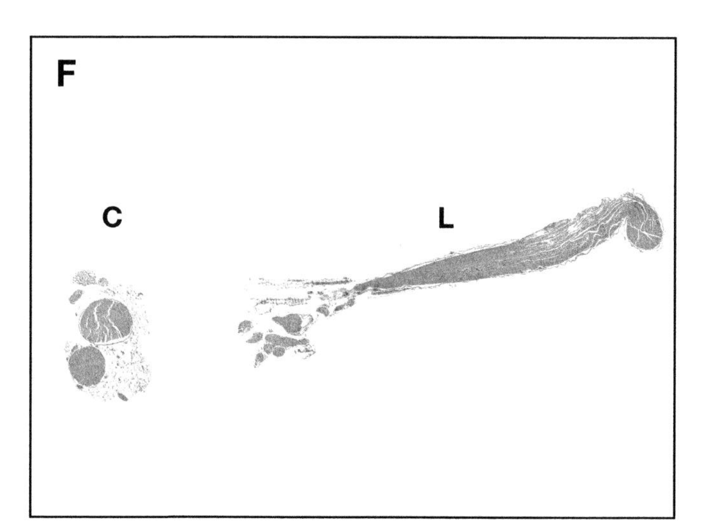

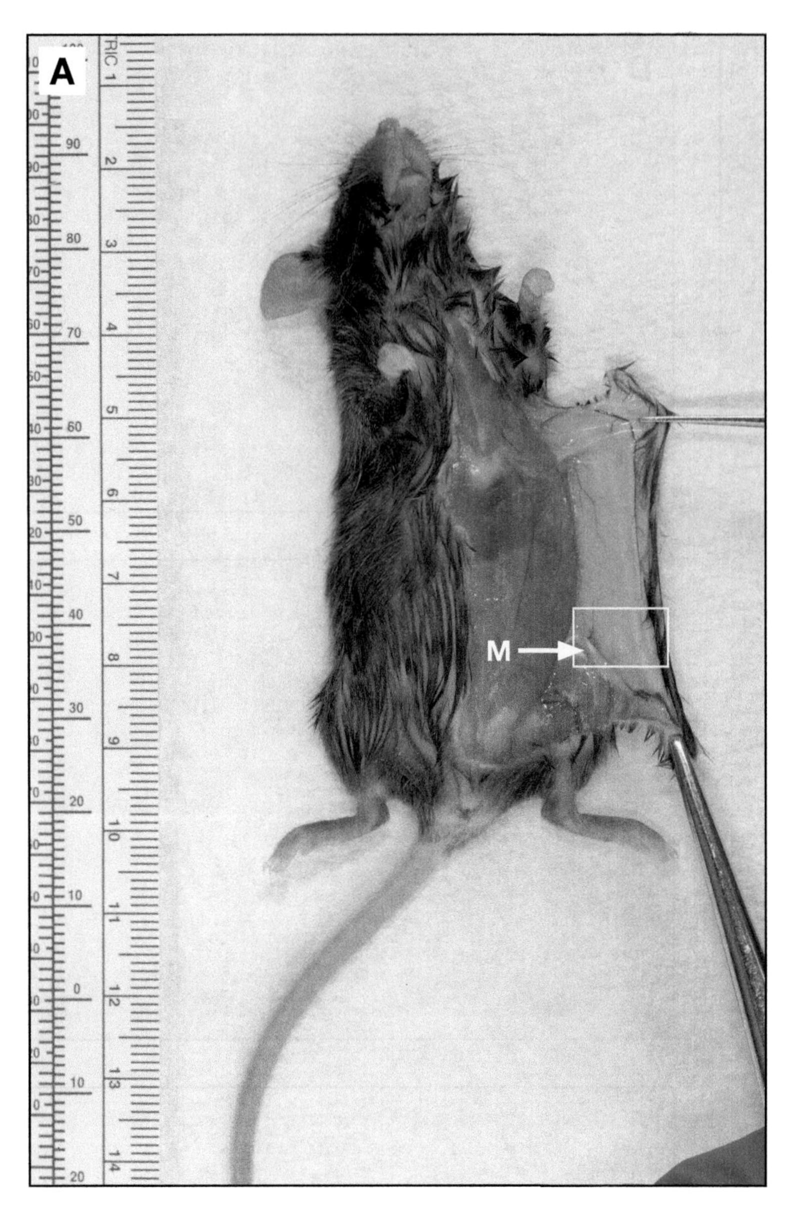

Skin with (or without) Mammary Gland

A. To collect both mammary gland (M) and skin, take a sample of skin from the inguinal region (box).

B. An approximately 1 cm - 2 cm square of tissue is enough for histology. To prevent the skin from curling during fixation and processing, lay it flat inside a folded piece of paper towel at necropsy.

C. After processing cut the sample in half to obtain two cross sections through the skin and mammary gland.

D. Embed the pieces diagonally, standing on the cut face so that a complete cross section can be taken through each piece of skin.

E. Face the block until all layers of both pieces of skin and mammary gland are exposed.

F. The ideal section contains mammary gland (M), dermis (D), epidermis (E) and fur (F).

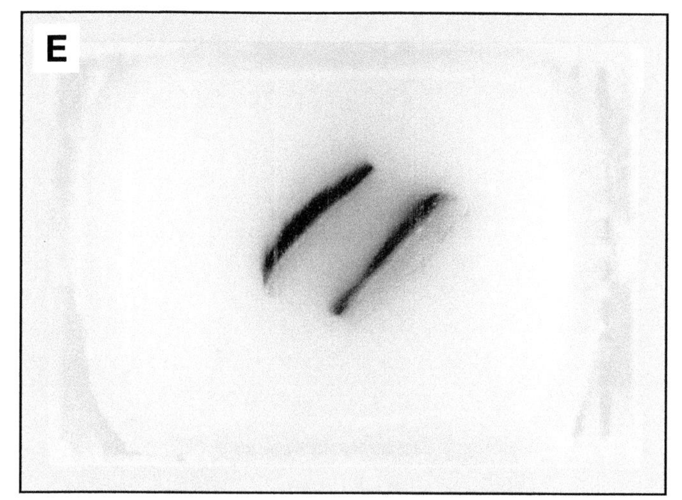

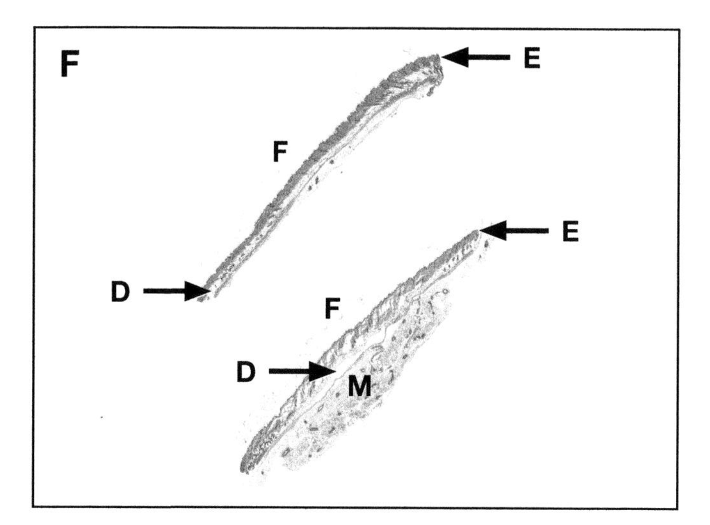

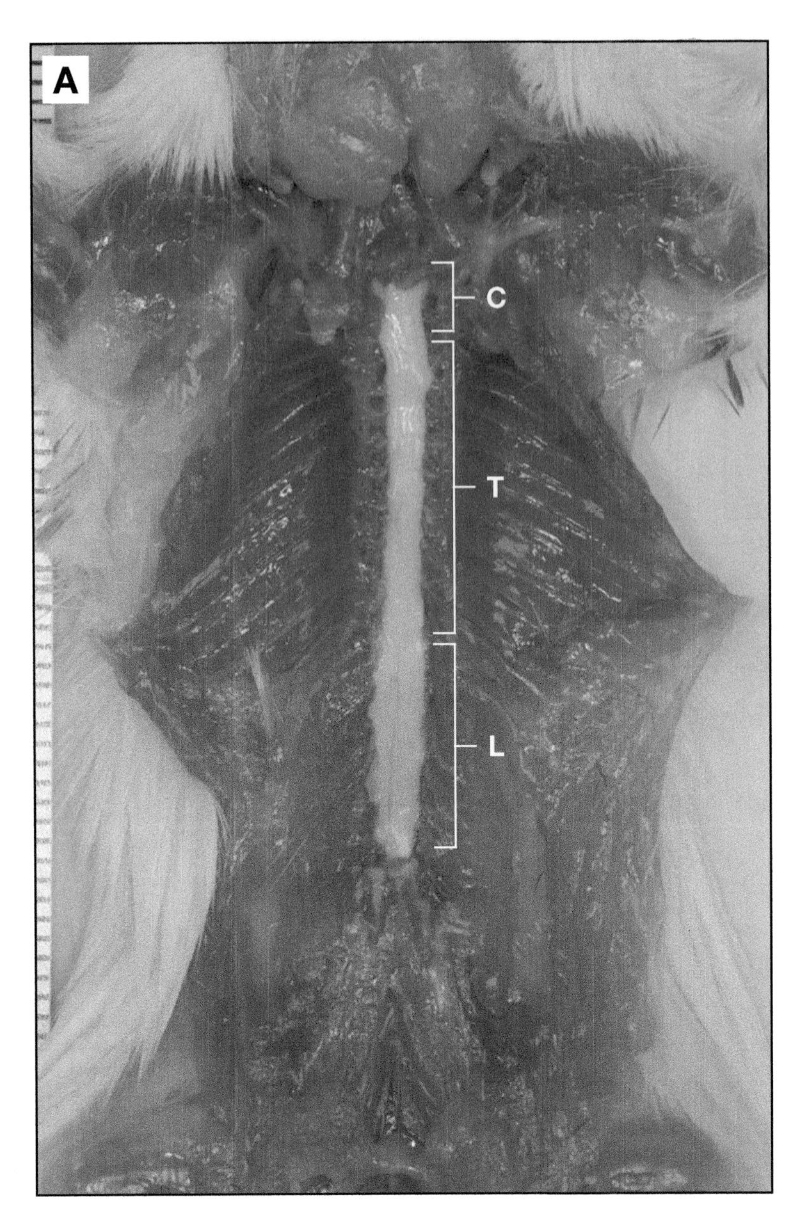

Spinal Cord

A. With the visceral organs removed, the ventral side of the spinal column is exposed. Remove these bones i.e., vertebral bodies covering the spinal cord using small sharp scissors. Areas of the white spinal cord i.e. cervical (C), thoracic (T), and lumbar (L) are located to left of the brackets. Removing the cord from the vertebra is necessary if subsequent procedures like staining or immunohistochemistry are affected by decalcification. The spinal cord can be carefully removed at necropsy. However, it is recommended that the spinal cord remain in situ to fix prior to removal. After fixation, the delicate spinal cord is firmer and easier to remove. If dorsal root ganglia are needed, the spinal cord within the bony vertebrae should be fixed, decalcified and processed, refer to Spine (page 55).

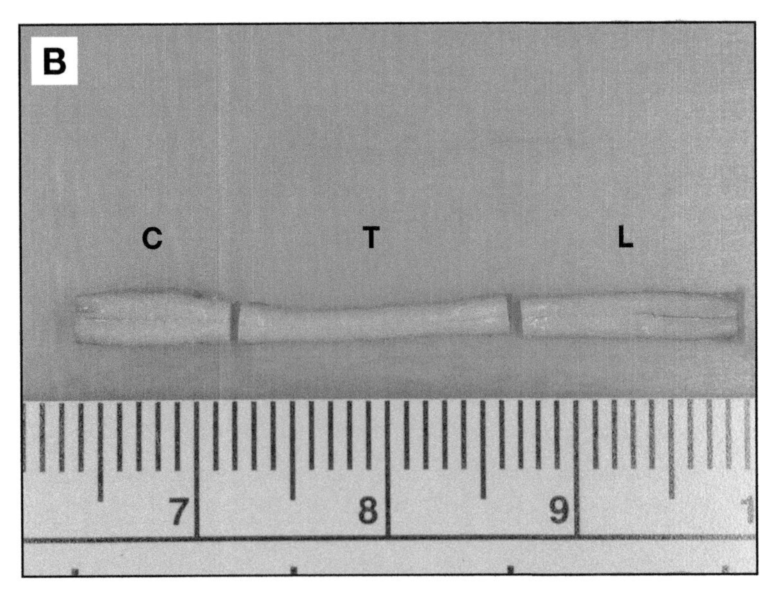

B. This photo shows the spinal cord trimmed into three regions i.e. cervical (C), thoracic (T), and lumbar (L).

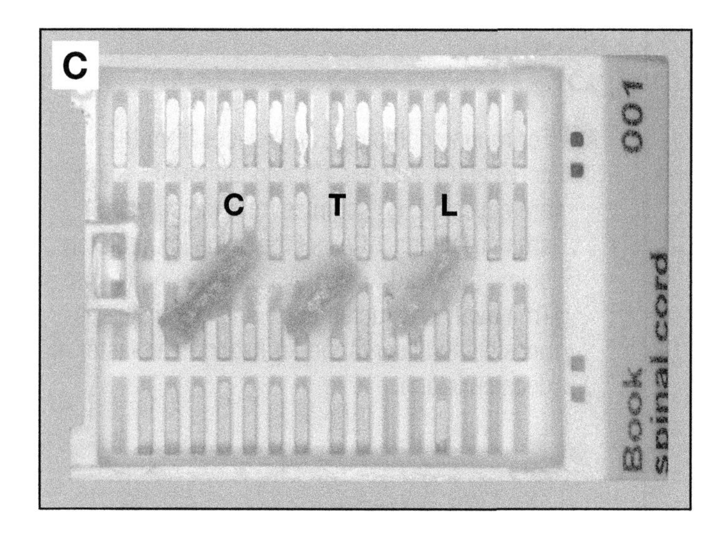

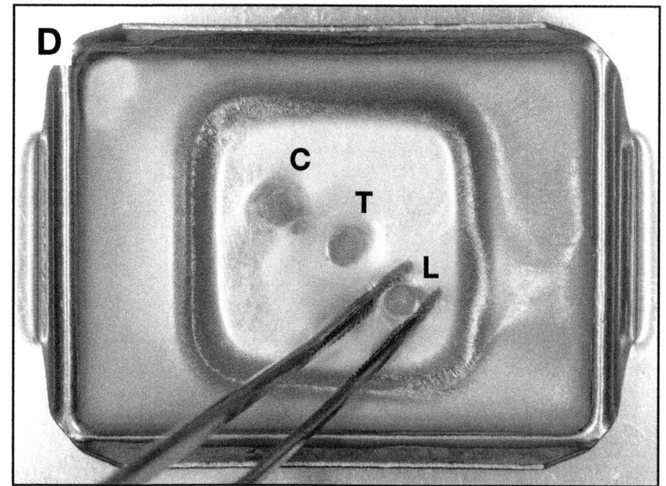

C. Pieces (1 cm - 2 cm) of cervical (C), thoracic (T), and lumbar (L) regions are processed together in a cassette.

D. All three pieces of cord are embedded on end to obtain cross sections.

E. Carefully face into the block until entire cross sections of all three pieces are obtained.

F. The cords are delicate and any slight mechanical damage will show on the final section of the cords.

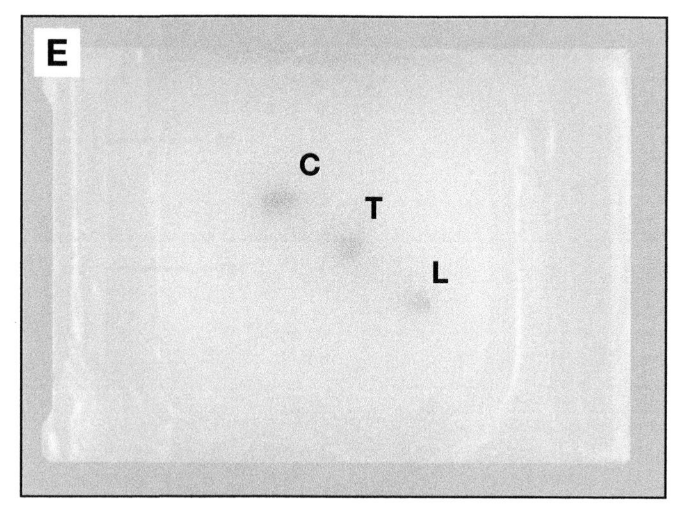

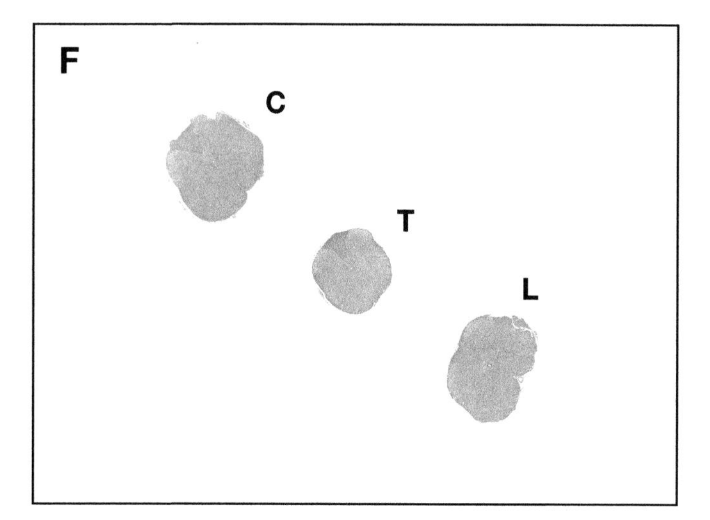

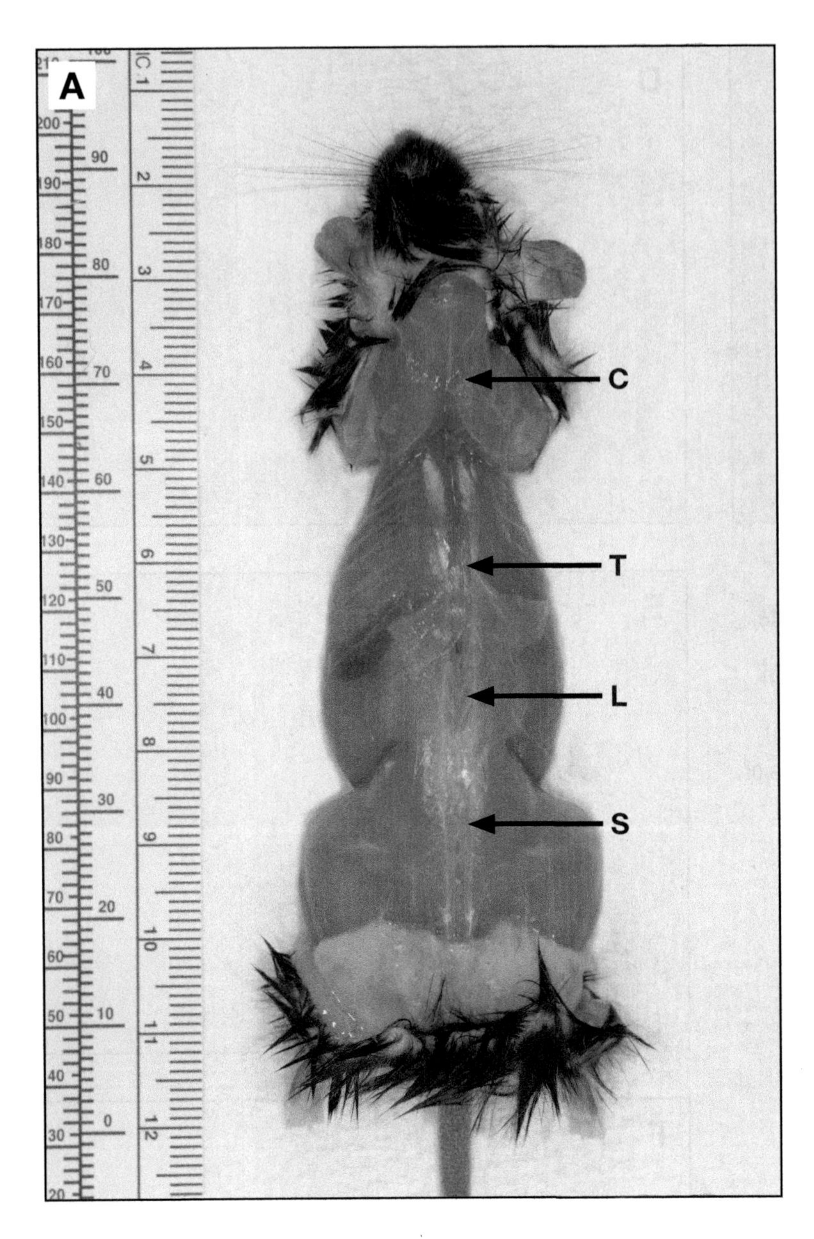

Spine

A. For routine pathology, only pieces from the cervical (C), thoracic (T), and lumbar (L) spine are collected; no samples from the sacral (S) area are needed. Collect the spine immediately below the head to slightly below the hip joints. Trim away as much muscle and soft tissue as possible to allow for better fixation and faster decalcification.

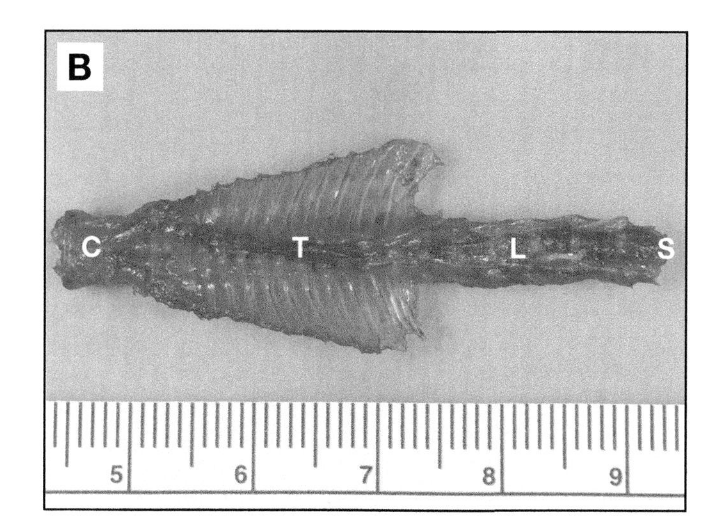

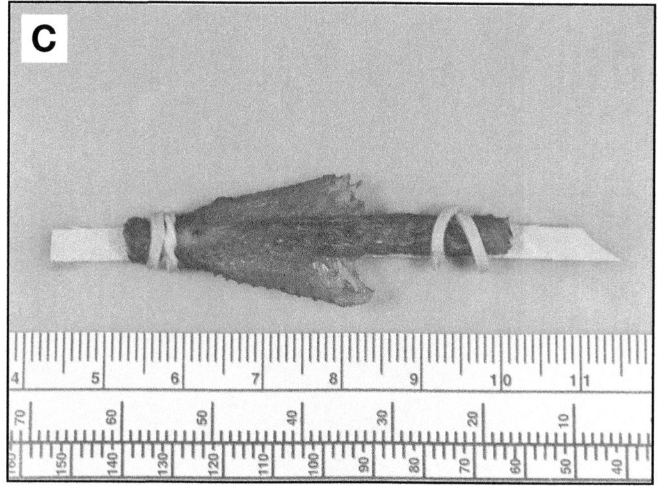

B. Cut the top of the spine from the head as close to the skull as possible to preserve cervical (C) vertebrae. Leave a few millimeters of ribs attached to the thoracic (T) spinal area to use as a landmark for later trimming. The lumbar (L) region begins below the ribs and ends at the longer bones of the sacral (S) vertebrae.

C. To hold the column straight during fixation and decalcification, use small elastic bands to bind the spine to a wooden coffee stirrer. The straight column is easier to separate into the specific regions.

Spine (Continued)

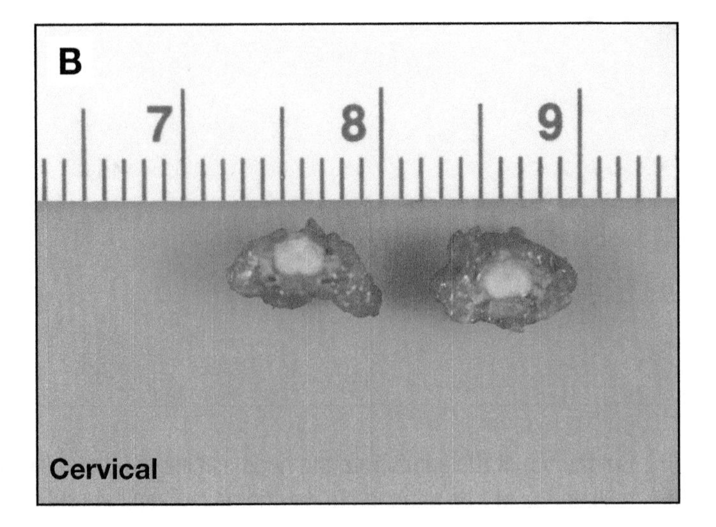

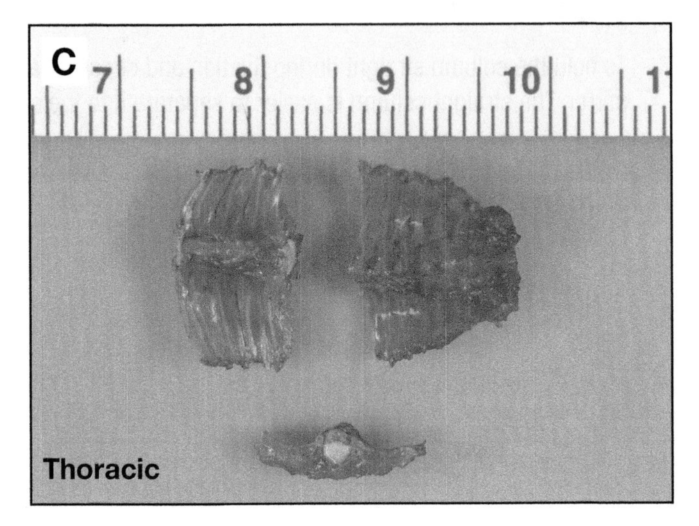

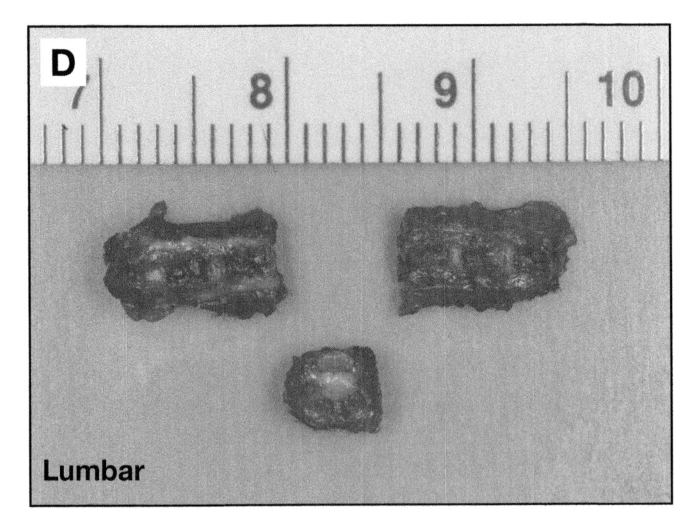

A. In this photo, the spine has been separated into three regions. The cervical region (C), beginning at the head and ending at the ribs, contains 7 vertebrae. The thoracic region (T) consists of the next 13 vertebrae with attached ribs. The lumbar region (L) consists of the next 6 vertebrae beginning immediately after the ribs and ending at the fused longer bone, the sacrum (S).

B. The cervical spine shown here is cut in half demonstrating cross sections of the white spinal cord (S).

C. A piece representing the thoracic spine is taken from the center of the thoracic region. A small length of rib is kept on the sample to help distinguish it from the other regions.

D. To ensure that the lumbar region is sampled, select a 1 cm piece from the center of the sample.

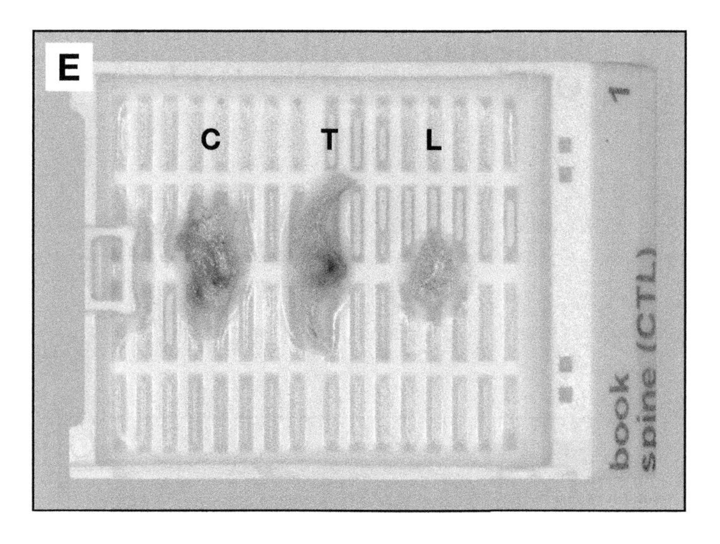

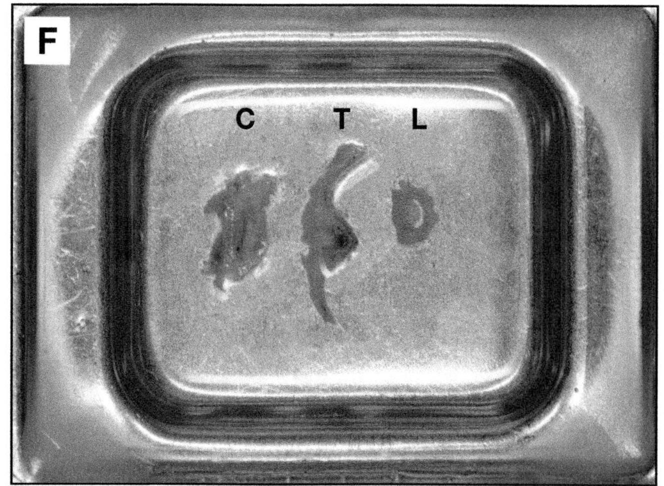

E. After decalcification and processing, the samples can appear more yellow to brown in color.

F. Embed the spine samples with the cleanest cut face down into the embedding mold with the thoracic piece in the center.

G. Carefully face into the block until all three regions of the spine are at full face.

H. The goal is a section containing both bony and spinal cord components from the cervical (C), thoracic (T) and lumbar (L) regions. Occasionally a dorsal root ganglion (D) is seen as in the cervical cross section.

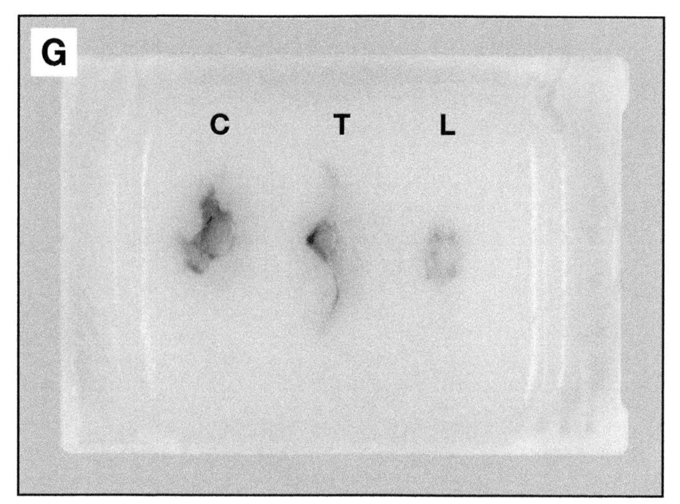

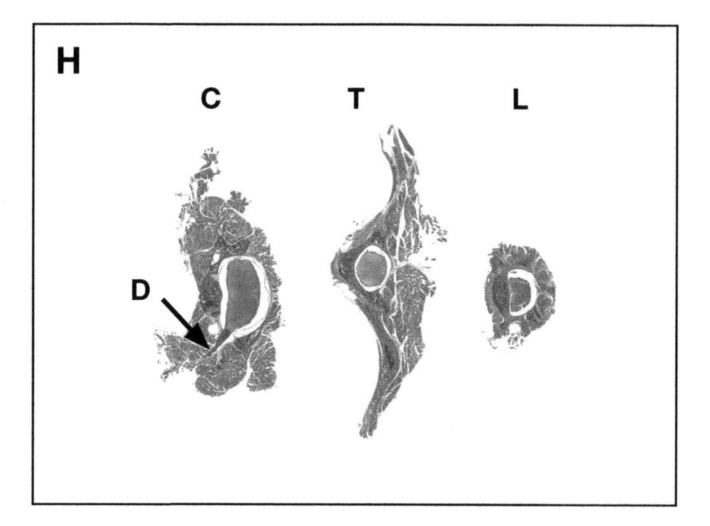

Spleen

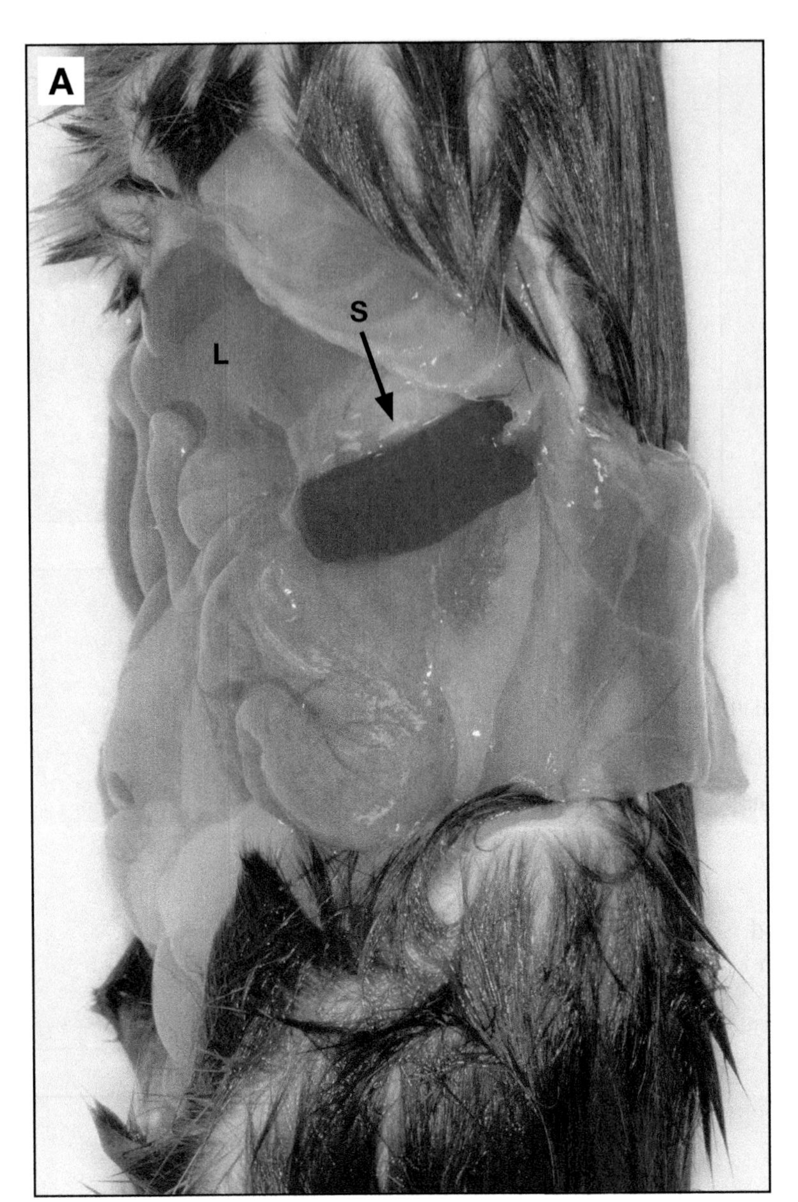

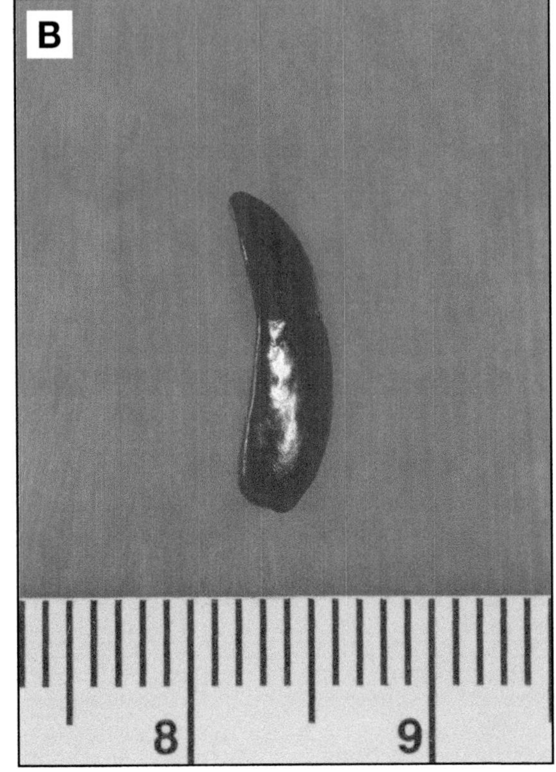

A. The spleen (S) is located on the left side of the abdominal cavity just below the liver (L). Remove the spleen by grasping the attached connective tissue in order to prevent any forceps artifacts.

B. This photo shows a whole spleen immediately after removal. The spleen can be cut transversely into three gross sections before fixation, processing, or embedding.

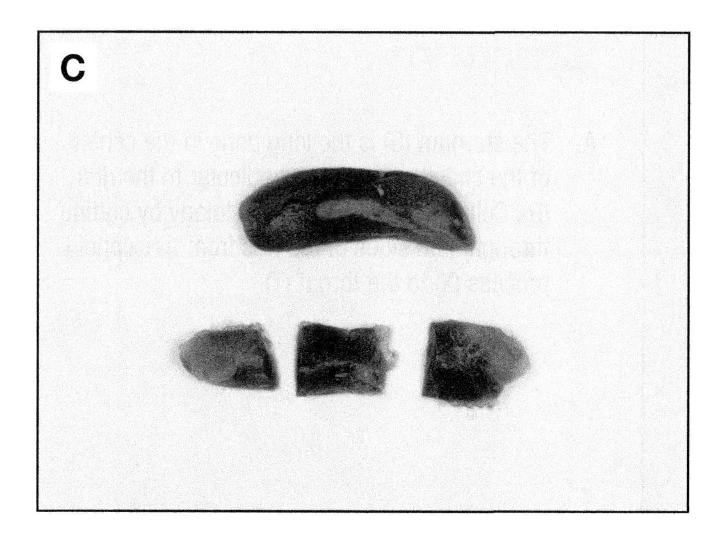

C. After processing, the spleen is darker red to brown in color. Prepare the spleen by cutting across the short edge.

D. Embed each piece on its cut face.

E. Face into the block until three complete pieces of spleen are seen.

F. This photo contains one of three triangular transverse sections of spleen. It is enlarged to show areas of white pulp (W) and red pulp (R).

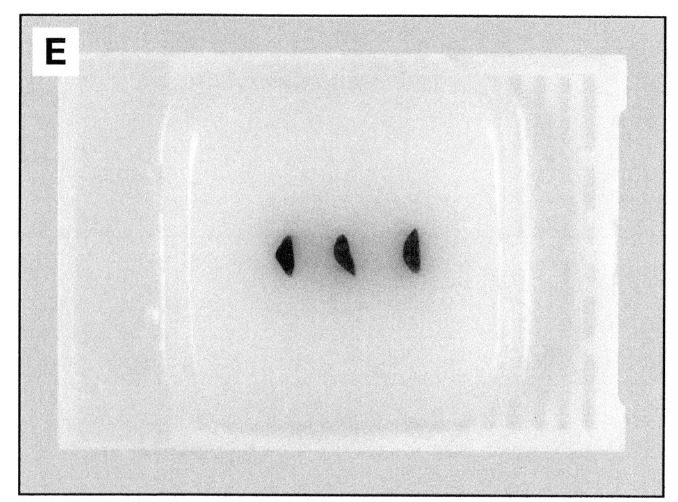

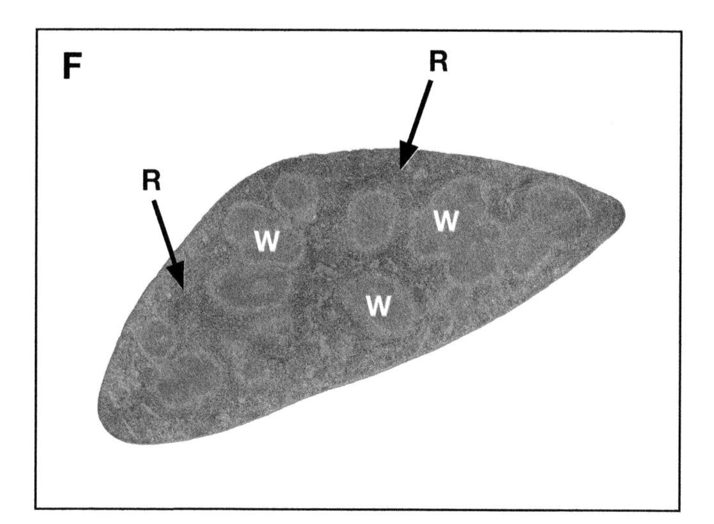

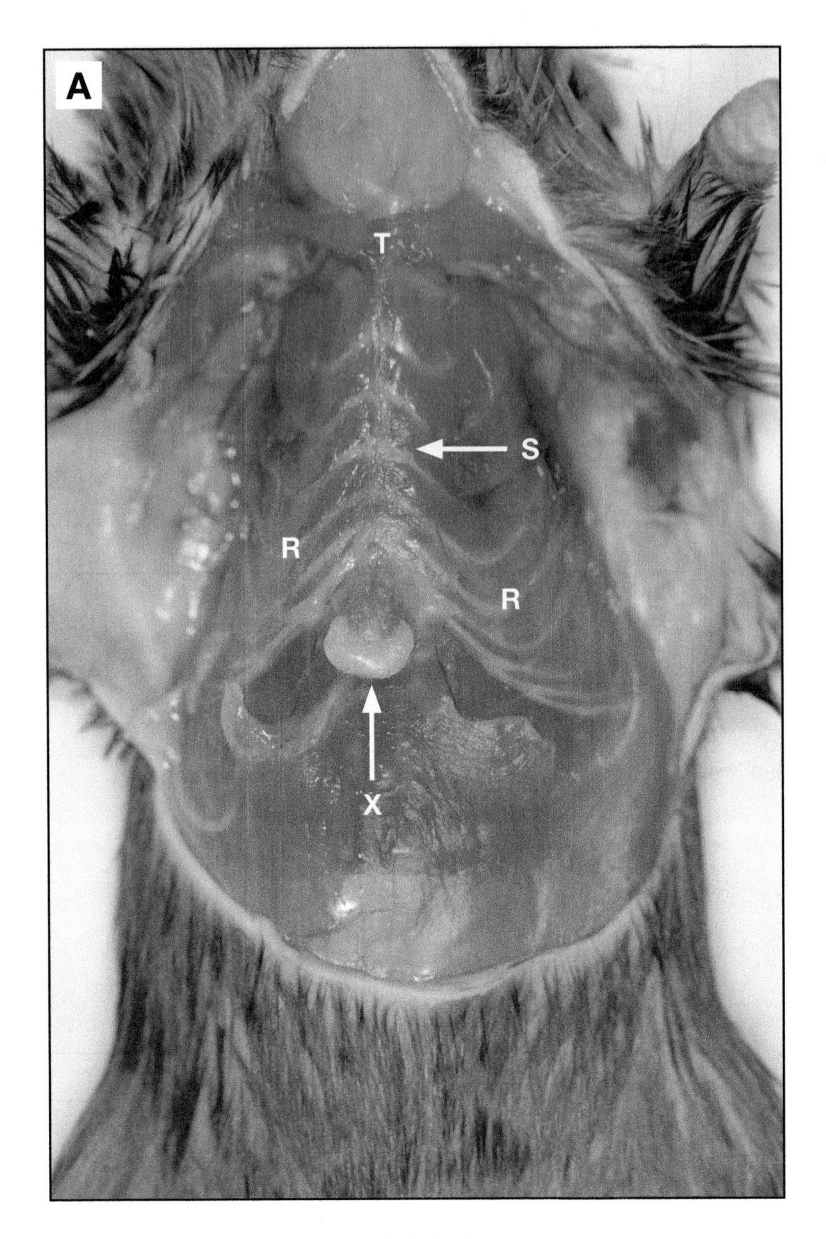

Sternum

A. The sternum (S) is the long bone in the center of the chest running perpendicular to the ribs (R). Collect the sternum for histology by cutting through both sides of the ribs from the xiphoid process (X) to the throat (T).

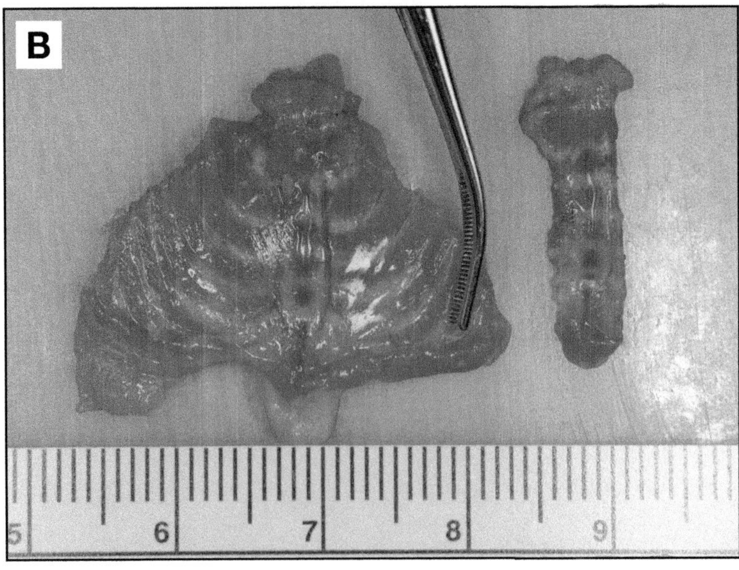

B. Once the ventral side of the rib cage has been removed from the body, more accurate trimming and removal of the ribs from the sternum can be performed. The sternum must be decalcified before paraffin processing.

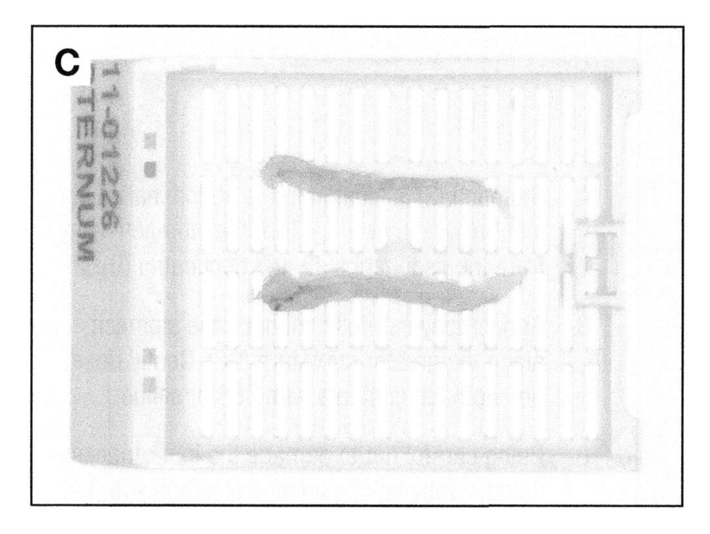

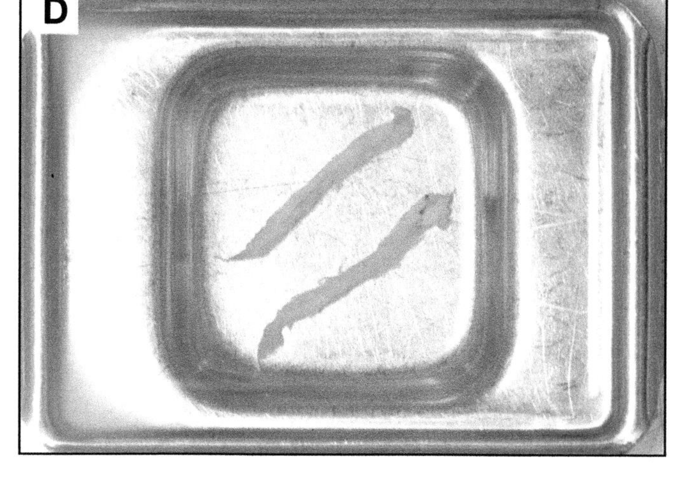

C. After decalcification, the sternum is trimmed in half lengthwise. After processing, the muscle and bone on these two long, strips are lighter in color. The bone marrow on the cut face (not shown here) will be a deep brick red.

D. Embed the cut faces of the sternum down into the mold. Press firmly to ensure the full lengths of the sternum and bone marrow are obtained.

E. Face into the block until the full length of the sternum is reached. Bone marrow (BM) is limited, so re-embed the sternum pieces if they are not lying flat.

F. The ideal section contains bone (B) and bone marrow (BM) along the entire length of the sternum.

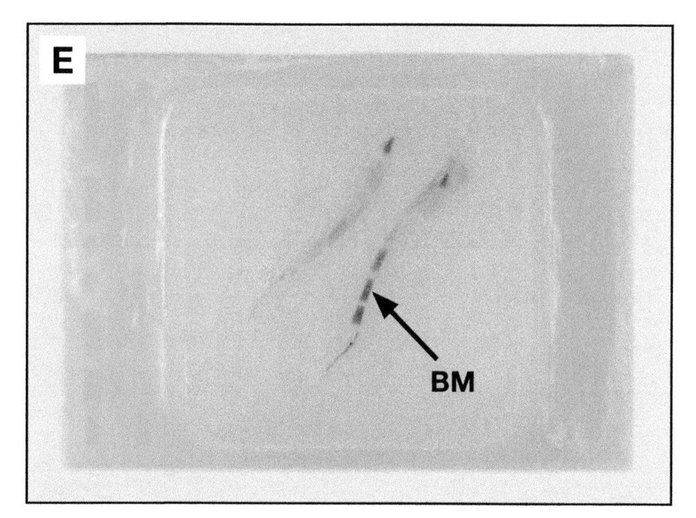

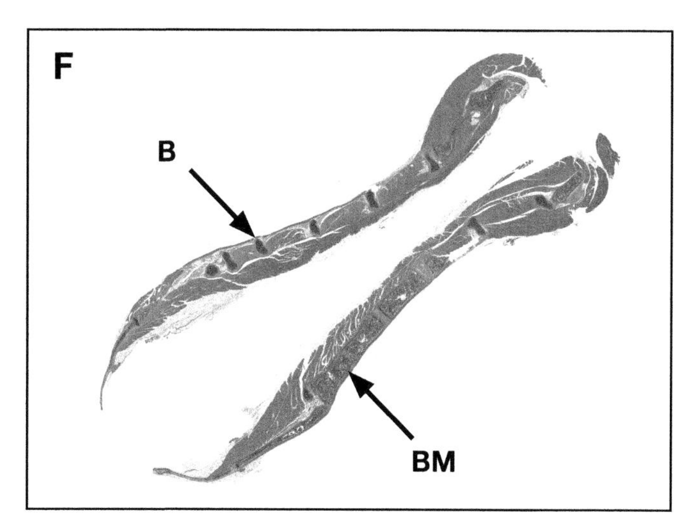

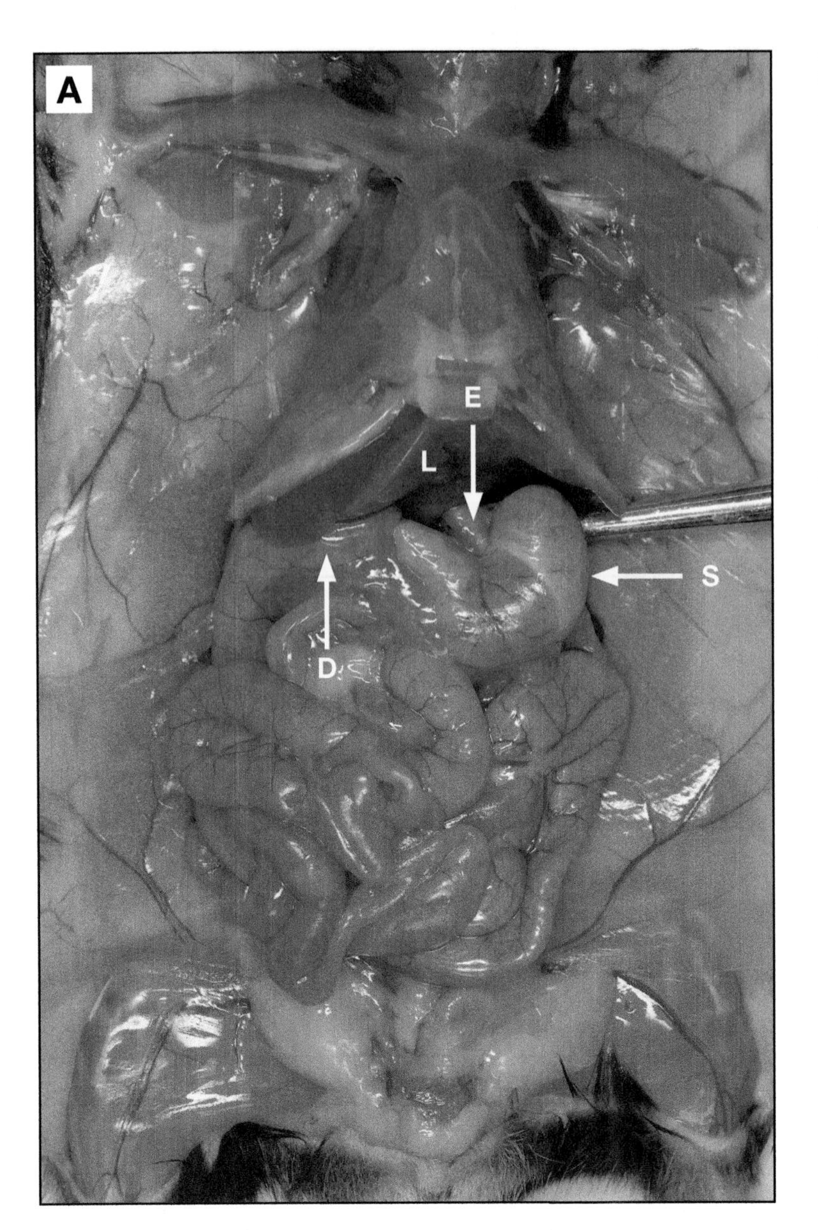

Stomach - Open Method

A. Stomach (S) is located in the abdominal cavity just below the liver (L). Trim the stomach away from the esophagus (E) and duodenum (D).

B. At necropsy carefully cut open the stomach along the greater curvature (GC). Gently rinse out stomach contents with PBS or saline.

C. Being careful not to damage the inner mucosal lining, gently lay the opened stomach with the mucosal surface facing down onto one edge of a piece of paper towel. Fold the other side of paper towel over to cover the stomach and help it lie flat when placed into a cassette.

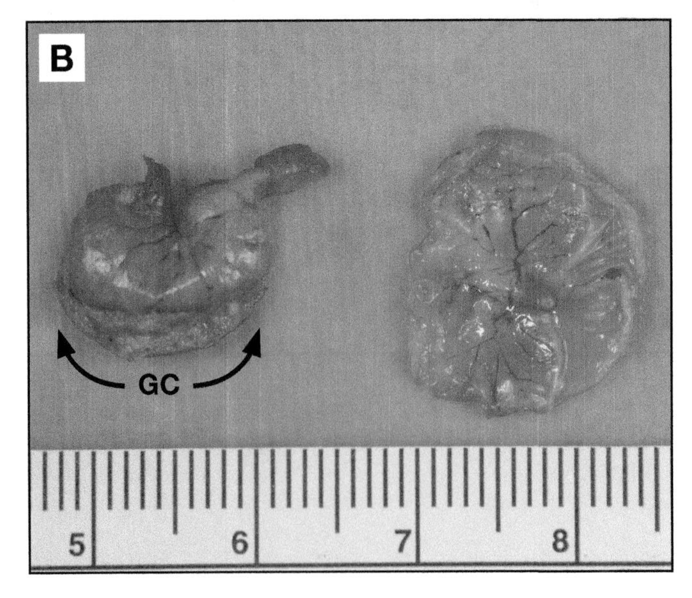

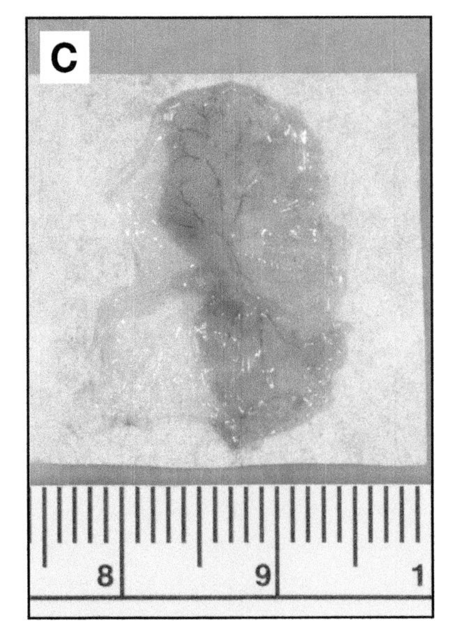

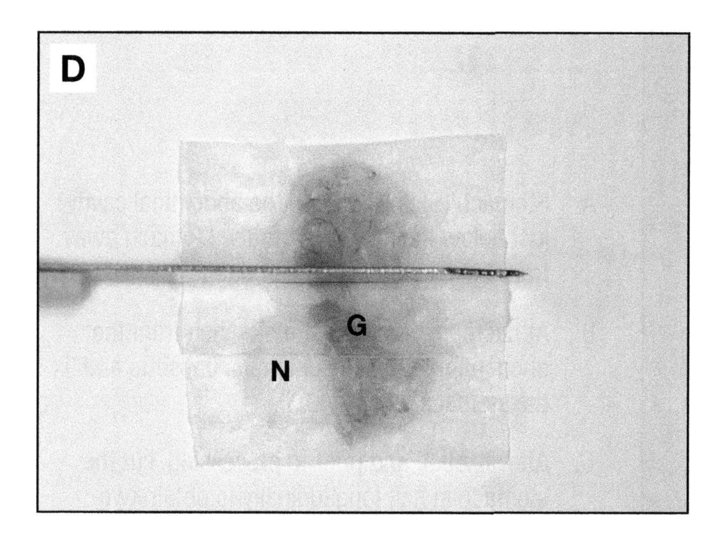

D. Determine where the duodenum and esophagus entered the stomach. Cut a 3 mm - 5 mm wide strip across the stomach that includes both landmarks. Be sure that the sample includes both the darker, thicker glandular area (G) and the translucent, thinner non-glandular area (N). Trimming may be done after fixation or as shown here, just prior to embedding.

E. Embed this strip of stomach on edge in order to get a cross section through the stomach wall. It may be necessary to hold this strip upright until the paraffin hardens at the bottom of the mold.

F. Face into the block until a full line of the stomach can be seen.

G. A proper section for histological analysis contains both glandular (G) and non-glandular (N) stomach.

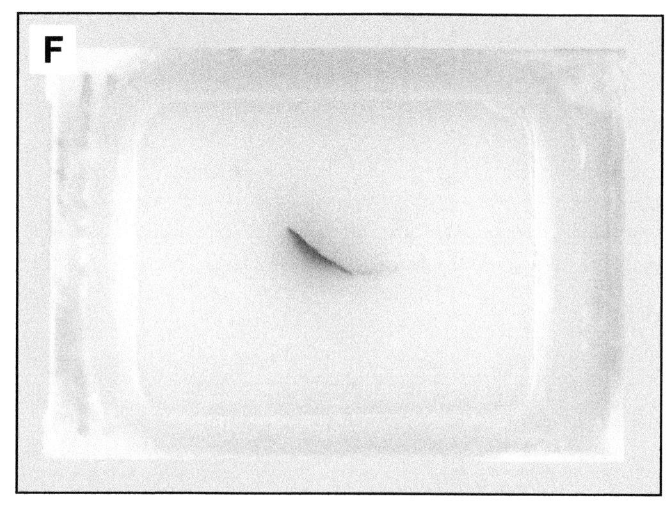

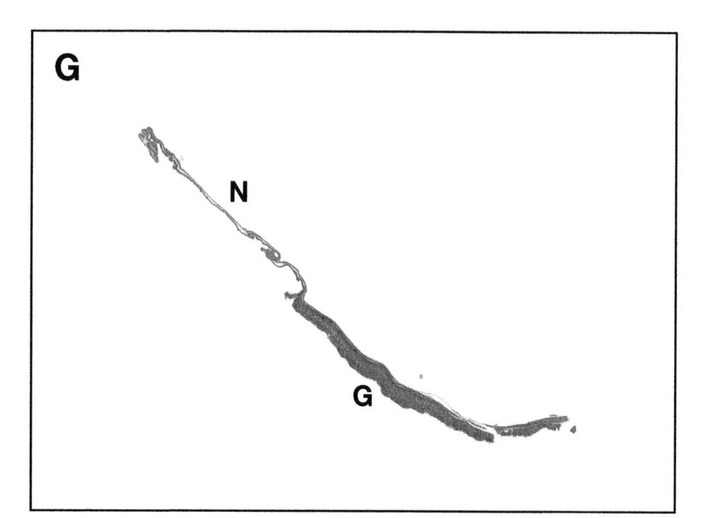

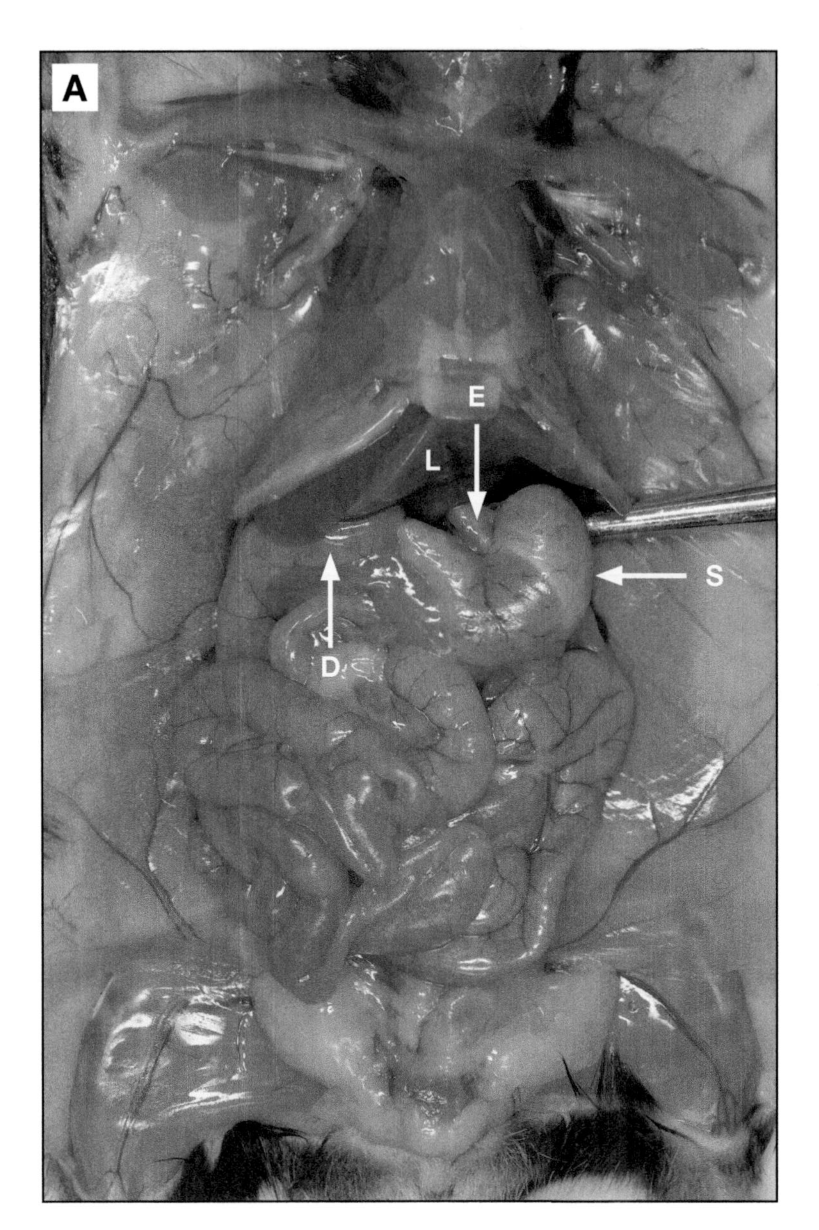

Stomach - Whole Method

A. Stomach (S) is located in the abdominal cavity just below the liver (L). Trim the stomach away from the esophagus (E) and duodenum (D).

B. At necropsy, separate the stomach from the duodenum and the esophagus. Cassette and fix the stomach whole.

C. After fixation and prior to processing, cut the stomach in half longitudinally to obtain two equal bean shaped pieces. Gently rinse out the stomach contents prior to re-cassetting the grossed stomach pieces.

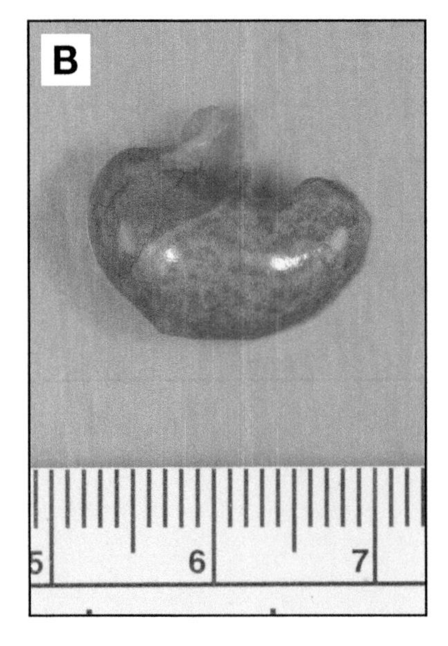

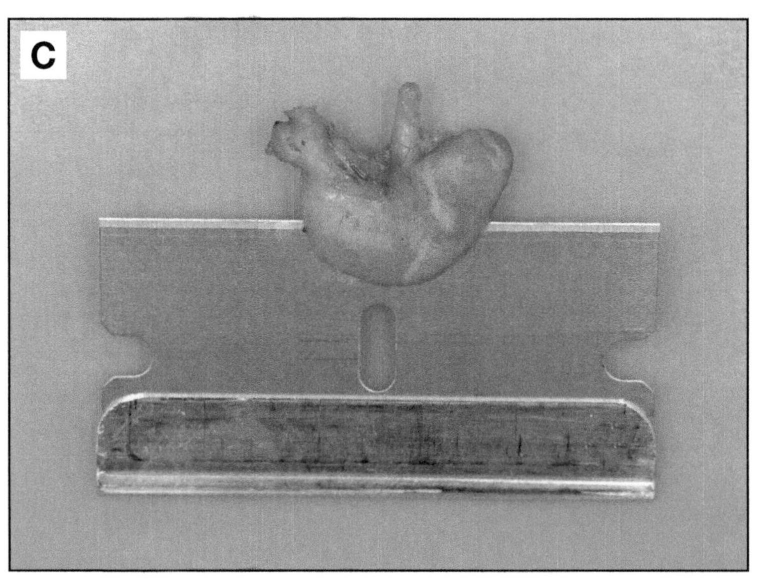

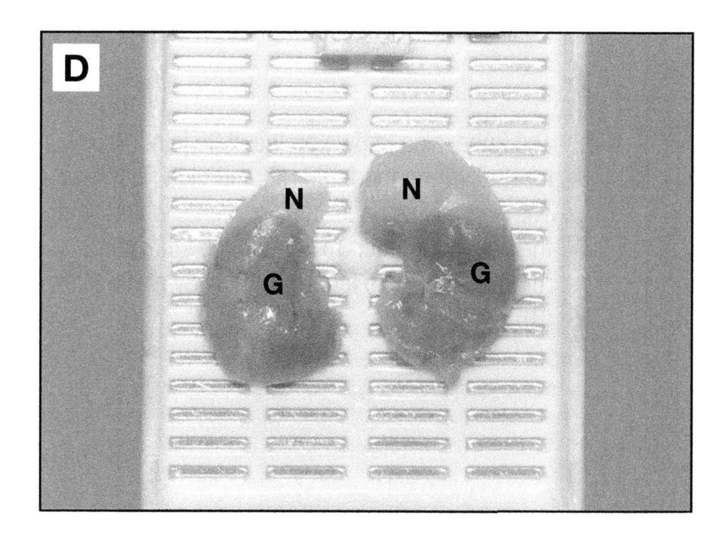

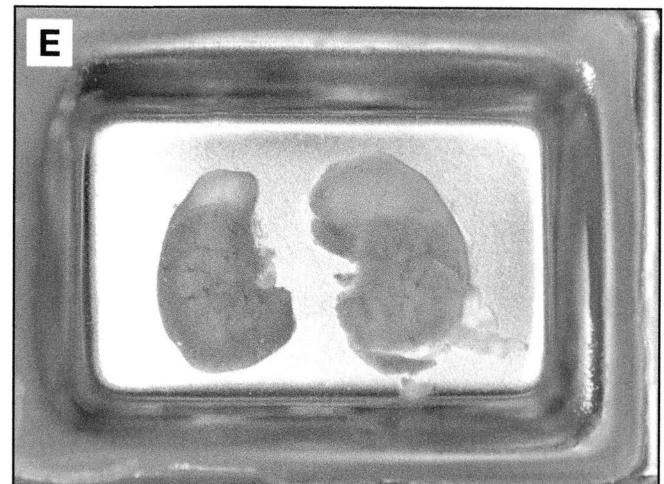

D. After processing, the glandular area (G) will be greyer in color and the non-glandular area (N) will be translucent.

E. Embed the two pieces of stomach with the open faces down into the mold.

F. Face into the block until the full outlines of both pieces of stomach are visible.

G. A good section will show a complete outline of the stomach with both the glandular (G) and non-glandular (N) areas on the slide. Small sections of pancreas (P) and esophagus (E) are present on this slide.

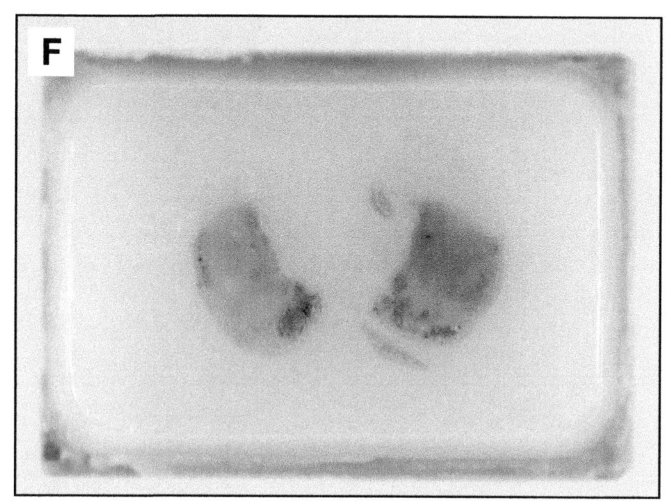

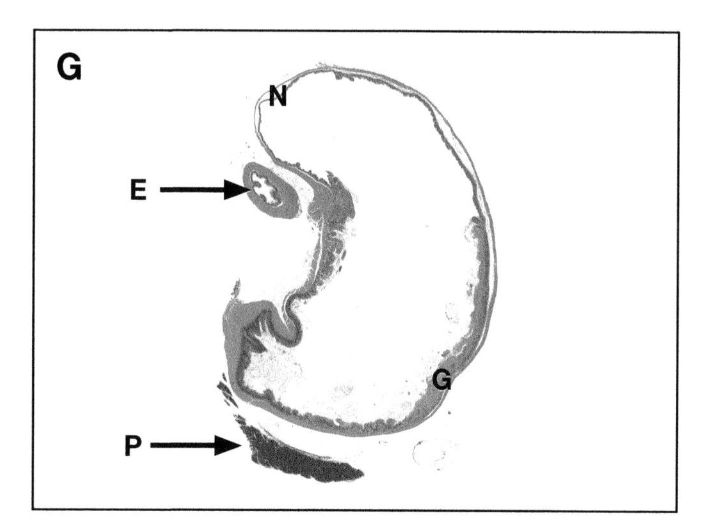

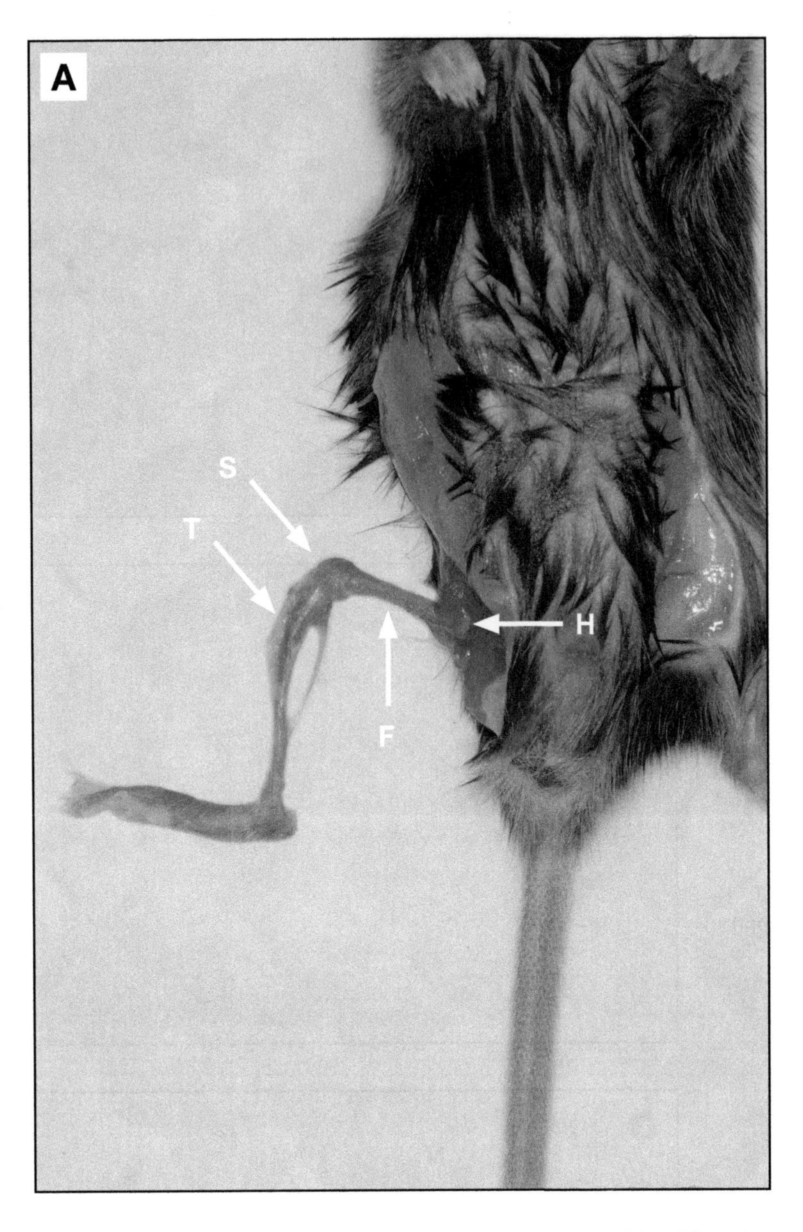

Stifle Joint

A. The knee, also known as the stifle joint (S), is located where the femur (F) and tibia (T) meet. The patella lies under the ligaments which hold the bones together. To sample the stifle joint, remove the whole leg from the body at hip joint (H).

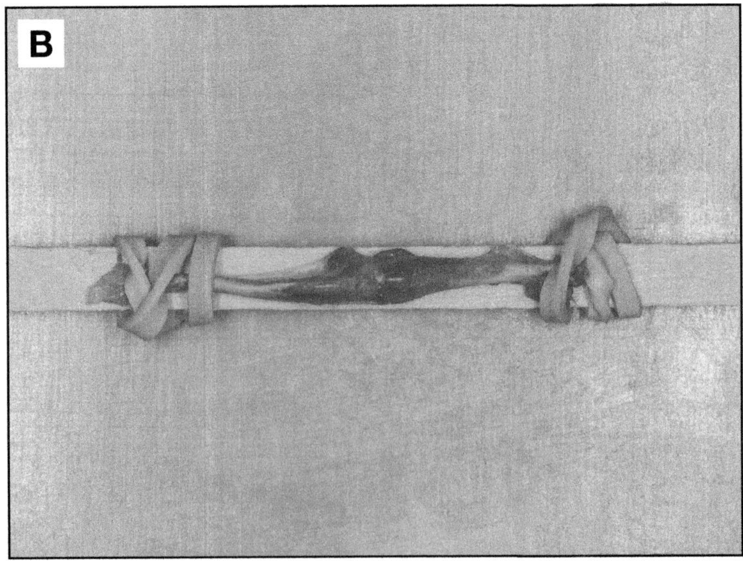

B. Once the leg is free from the body, the foot and all leg muscles are removed, the leg is "splinted" onto a wooden coffee stirrer using small elastic bands. This splinting holds the stifle joint straight during fixation, decalcification and processing in order to retain the best possible joint architecture.

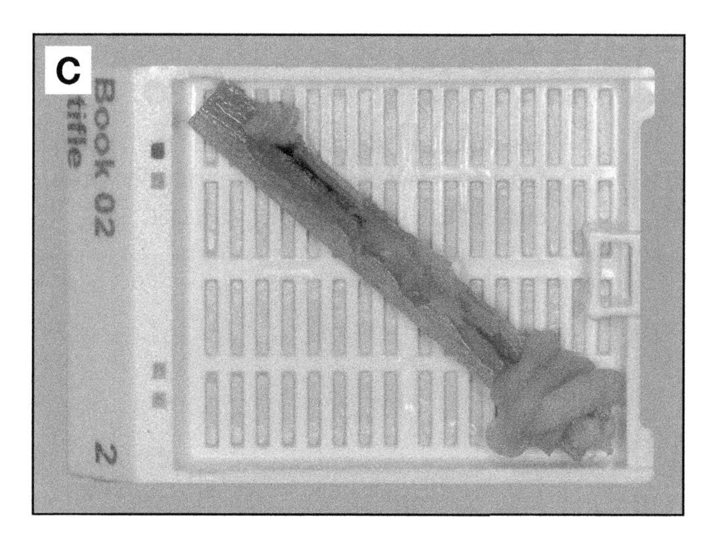

C. The stifle joint can either be removed from or processed on the splint. Here, the bone marrow within the bone has turned darker brown and the muscle that remained during processing is now tan in color.

D. Remove the splint and place the stifle joint with the patella side down and press firmly to ensure it is embedded squarely in the mold.

E. When facing into the block, it may be necessary to view sections microscopically to ensure the center of the joint is reached.

F. Correct sample orientation will show the following elements: the joint space (J), meniscus (M), growth plates (G), articular cartilage (C), cortical bone (B) and bone marrow (BM).

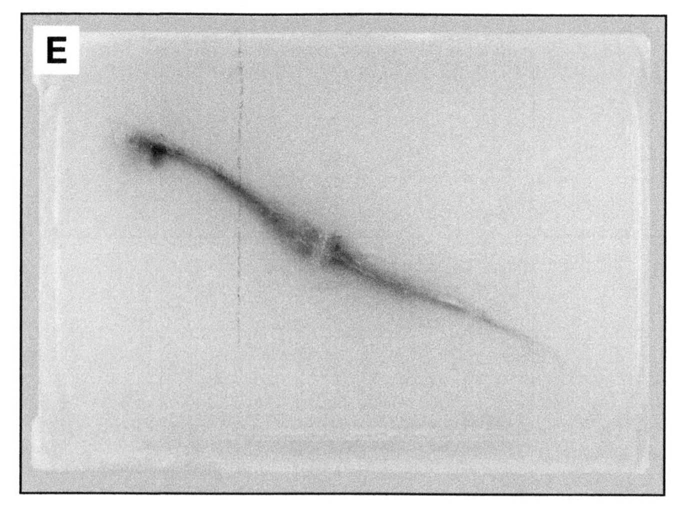

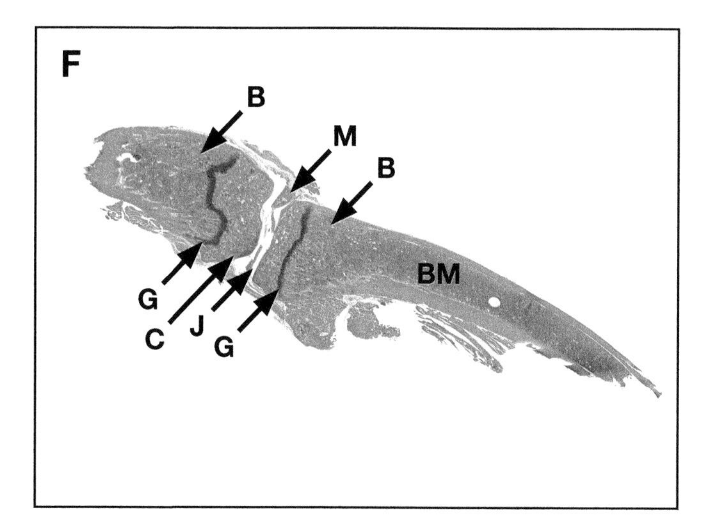

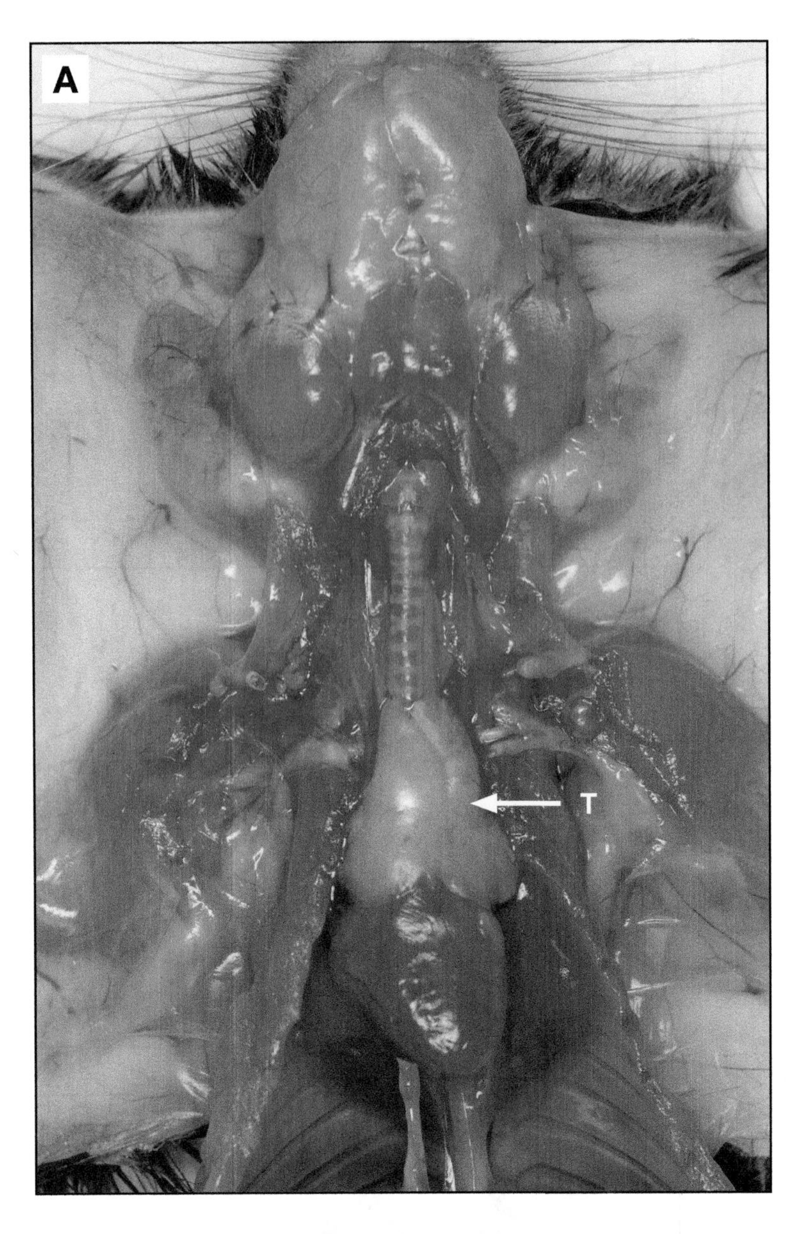

Thymus

A. The thymus (T) is located on top of the heart at the cranial end of the thoracic cavity. The lobes should be removed together, as a unit.

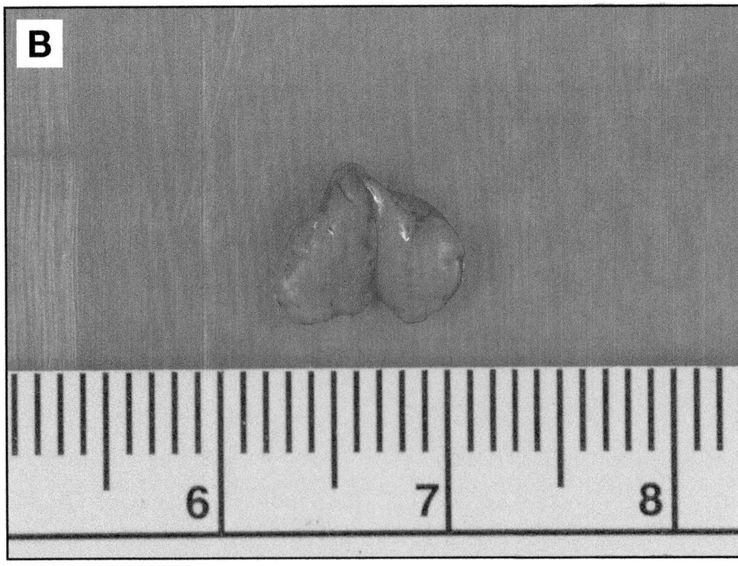

B. The thymus can be cassetted alone. If it is unusually small, use a CellSafe™ Biopsy Insert or piece of paper towel to prevent it from being lost during processing.

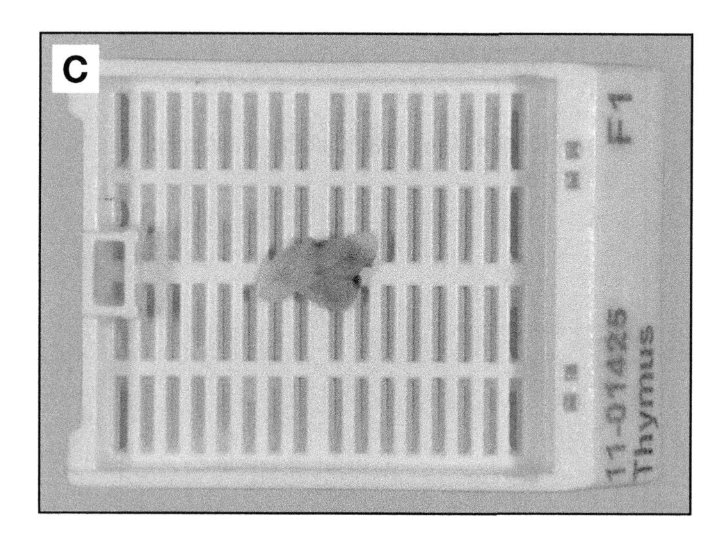

C. After fixation and processing, the thymus appears smaller and darker in color.

D. The ventral face of the thymus should be embedded with the rounded side down so that both lobes are seen in a section.

E. Face into the block until the entire surface of the thymus can be seen.

F. The ideal section will contain both lobes of the thymus.

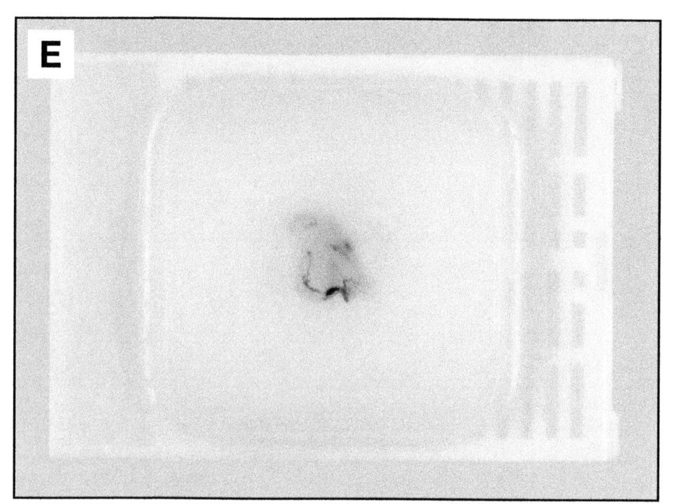

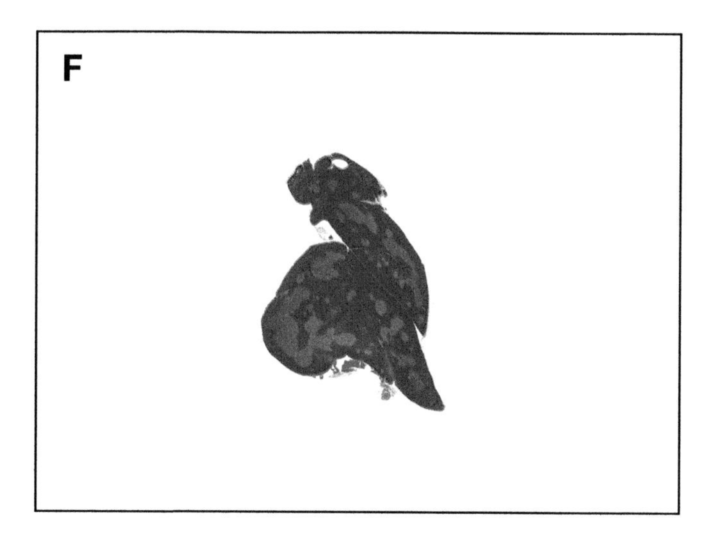

Tongue

A. Tongue (T) is located in the oral cavity.

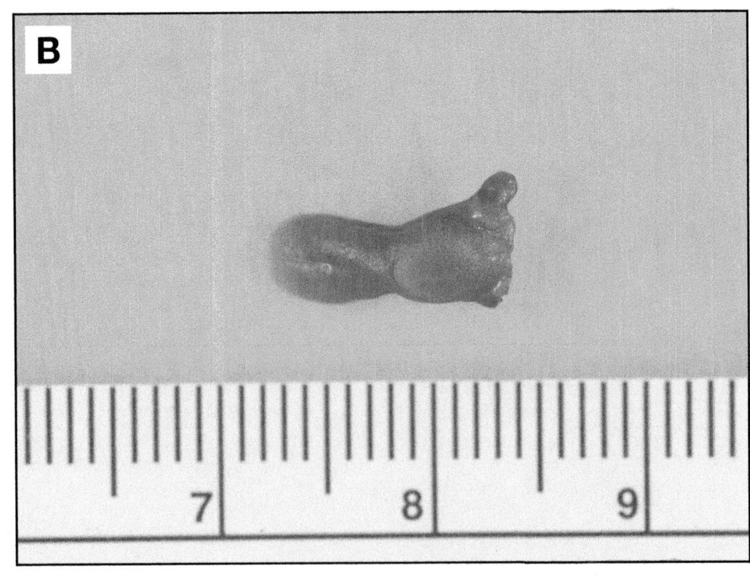

B. Remove the tongue whole when collecting for histology.

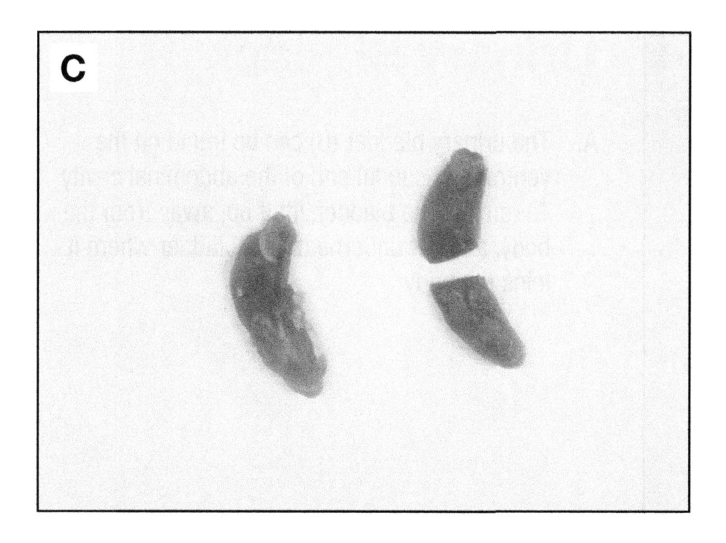

C. After processing, the tongue is cut transversely, along the short edge.

D. Embed both cut faces down into the embedding mold. If necessary, hold the pieces upright as the paraffin hardens.

E. Face into the block until a full cross section of both pieces of tongue can be seen.

F. An ideal section shows tongue epithelium (E) and underlying skeletal muscle fibers (M).

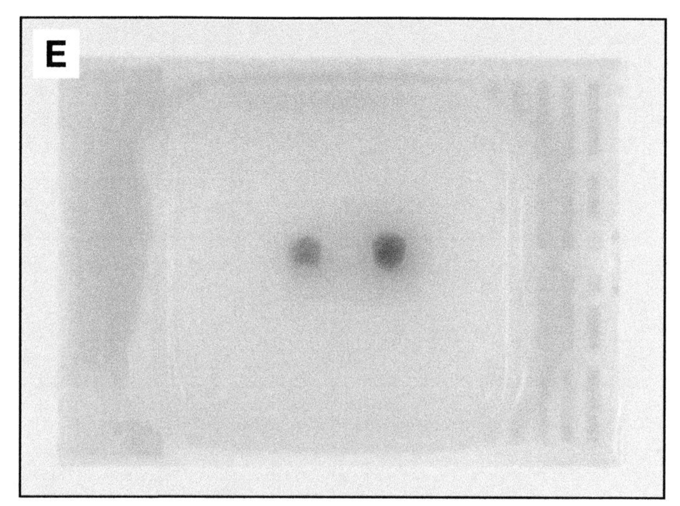

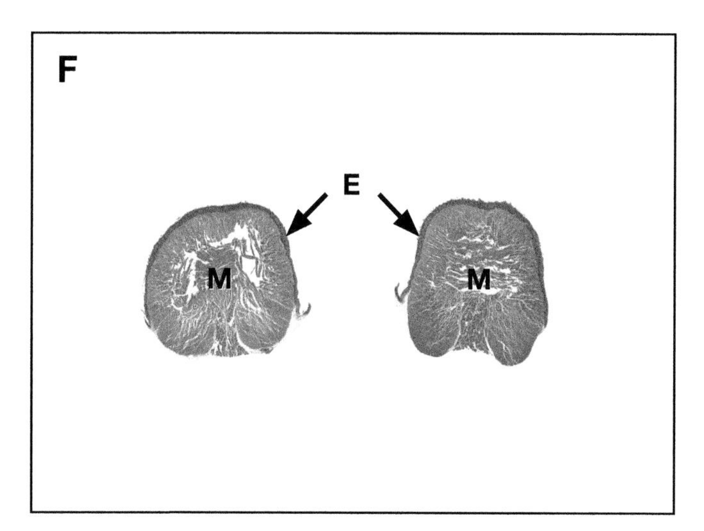

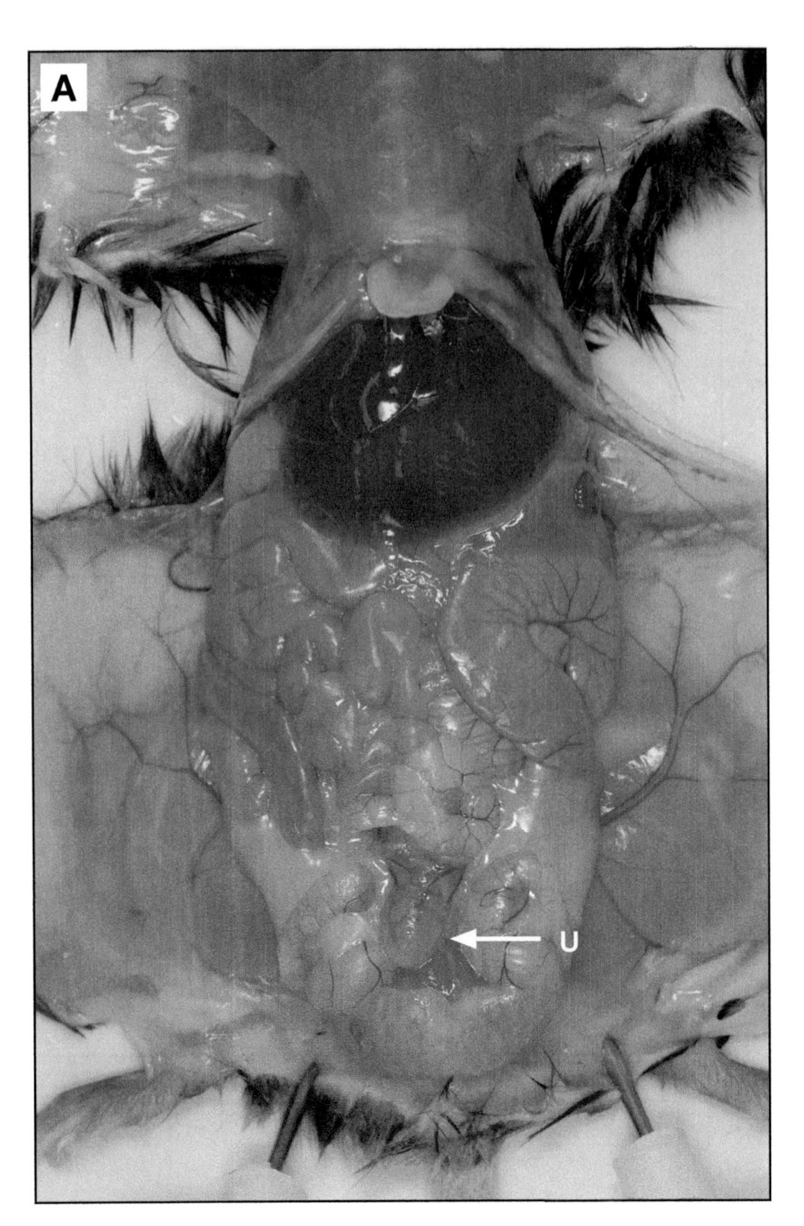

Urinary Bladder

A. The urinary bladder (U) can be found on the ventral and caudal end of the abdominal cavity. To remove the bladder, lift it up, away from the body, and cut underneath the bladder where it joins the body.

B. At necropsy, the bladder is sometimes filled with urine (shown on right). If the bladder is empty (left), place it into a CellSafe™ Biopsy Insert or wrap in a piece of paper towel to secure.

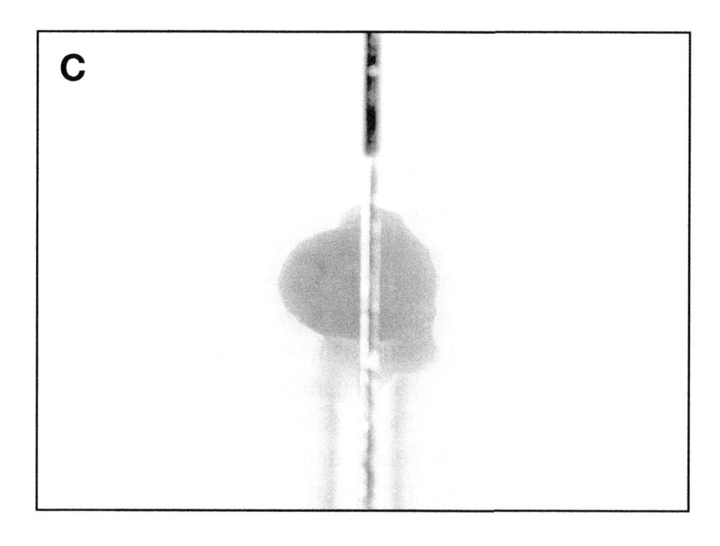

C. After processing, this empty bladder is smaller and greyer in color. Cut the bladder to obtain two equal cross sections.

D. To embed the urinary bladder, try to identify and place the cut surfaces down into the embedding mold. It may be necessary to hold them upright until the paraffin hardens at the base of the mold.

E. Carefully face into the cross sections because they are very small tissues.

F. A good section of the bladder contains a complete cross section through the bladder walls (W) and mucosa (M).

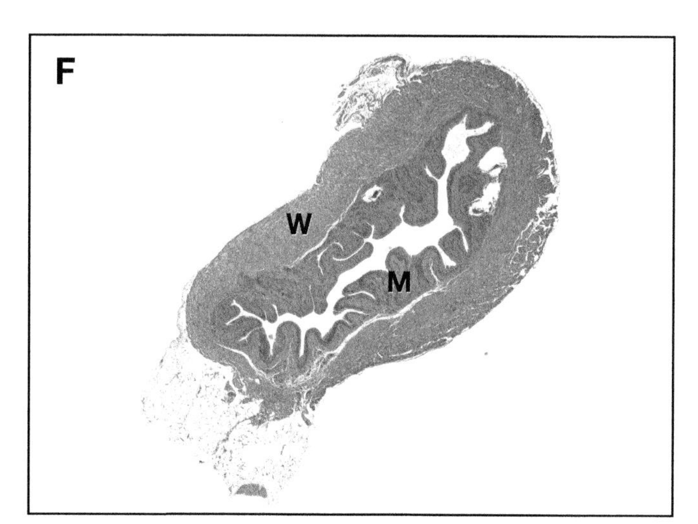

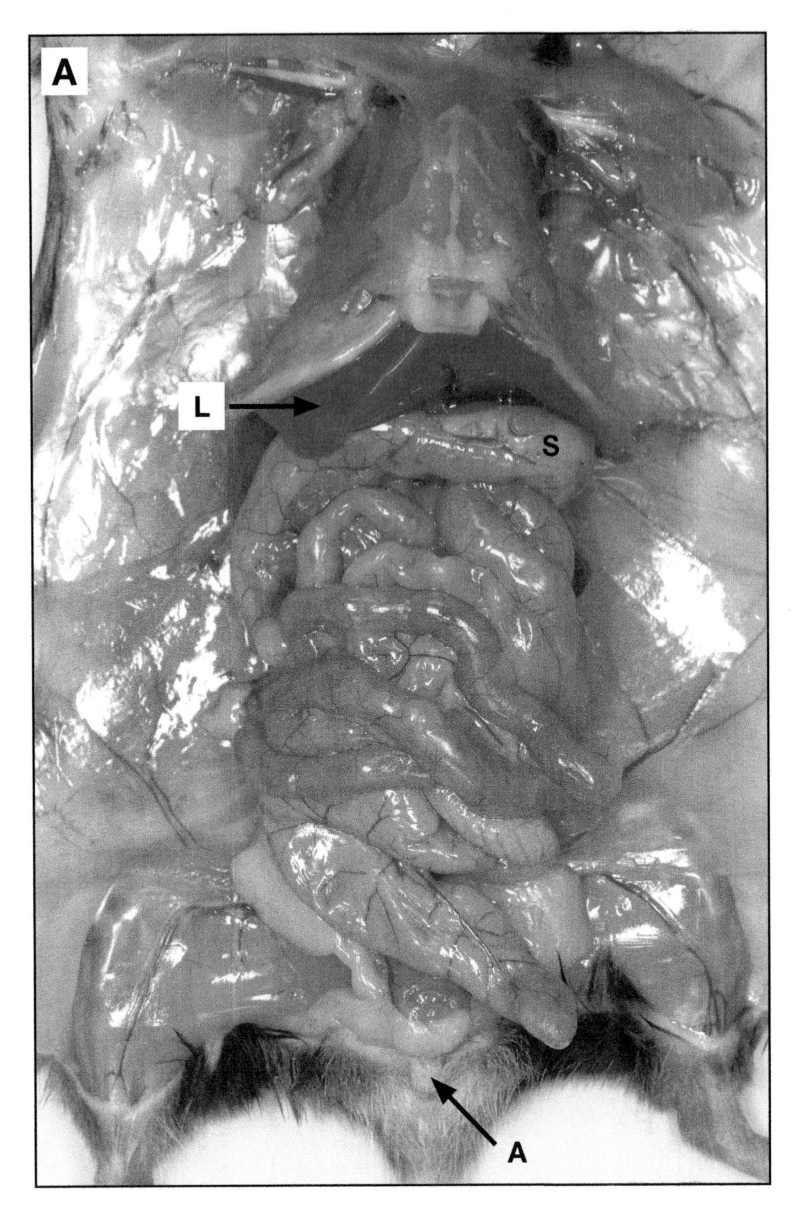

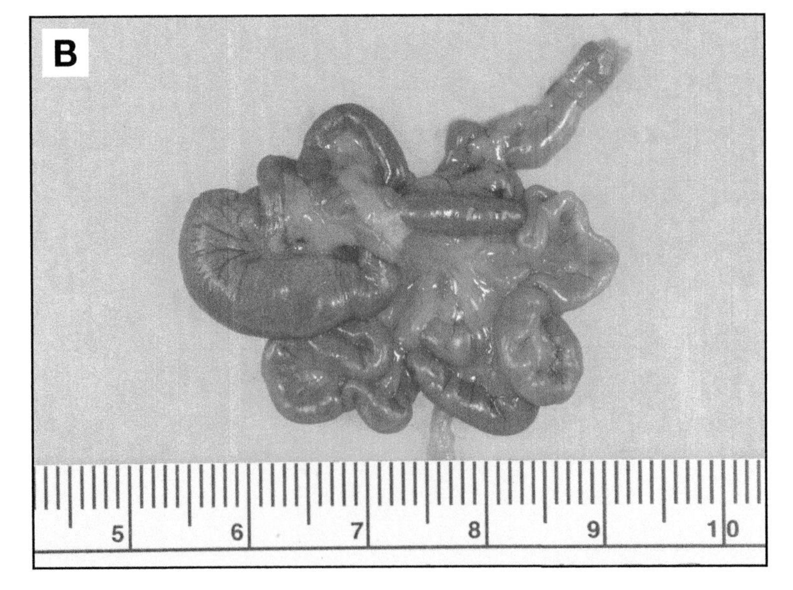

Intestines

A. Intestines are located in the abdominal cavity. They can be collected in two different ways – either as a whole unit or by individual regions. To remove them as a unit, trim the intestines away from the stomach (S) and liver (L) by slowly pulling the intestines forward and caudally. Cut away any organs (pancreas, spleen not visible in this photo) and ligaments connecting the intestines to the dorsal body wall. Carefully cut through the pelvic bone to finish dissecting the intestines out to the anus (A).

B. Included in this intestinal complex are small and large intestines, mesentery, mesenteric lymph nodes and abdominal fat.

C. Intestines may be flushed with fixative to both remove the fecal contents and reduce autolytic enzyme artifacts prior to immersion fixation. It may be easier to distinguish mesenteric lymph nodes from fat and mesentery after the intestines are fixed.

D. Stomach and intestinal tract with mesentery removed.

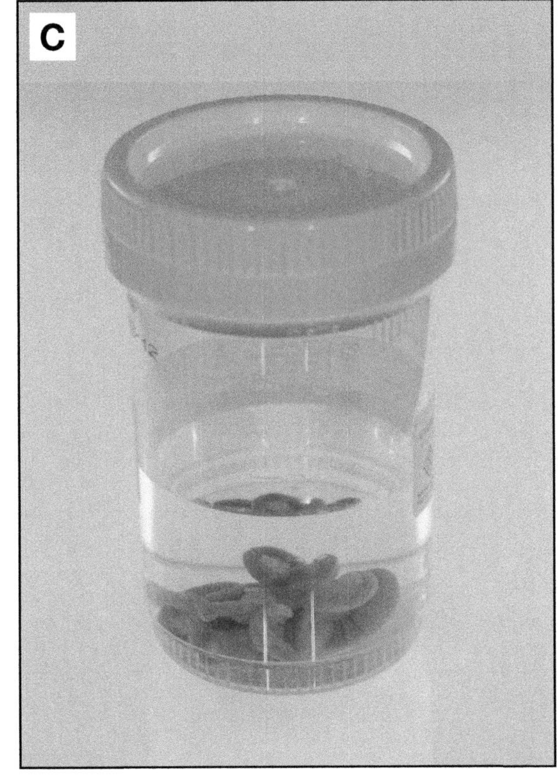

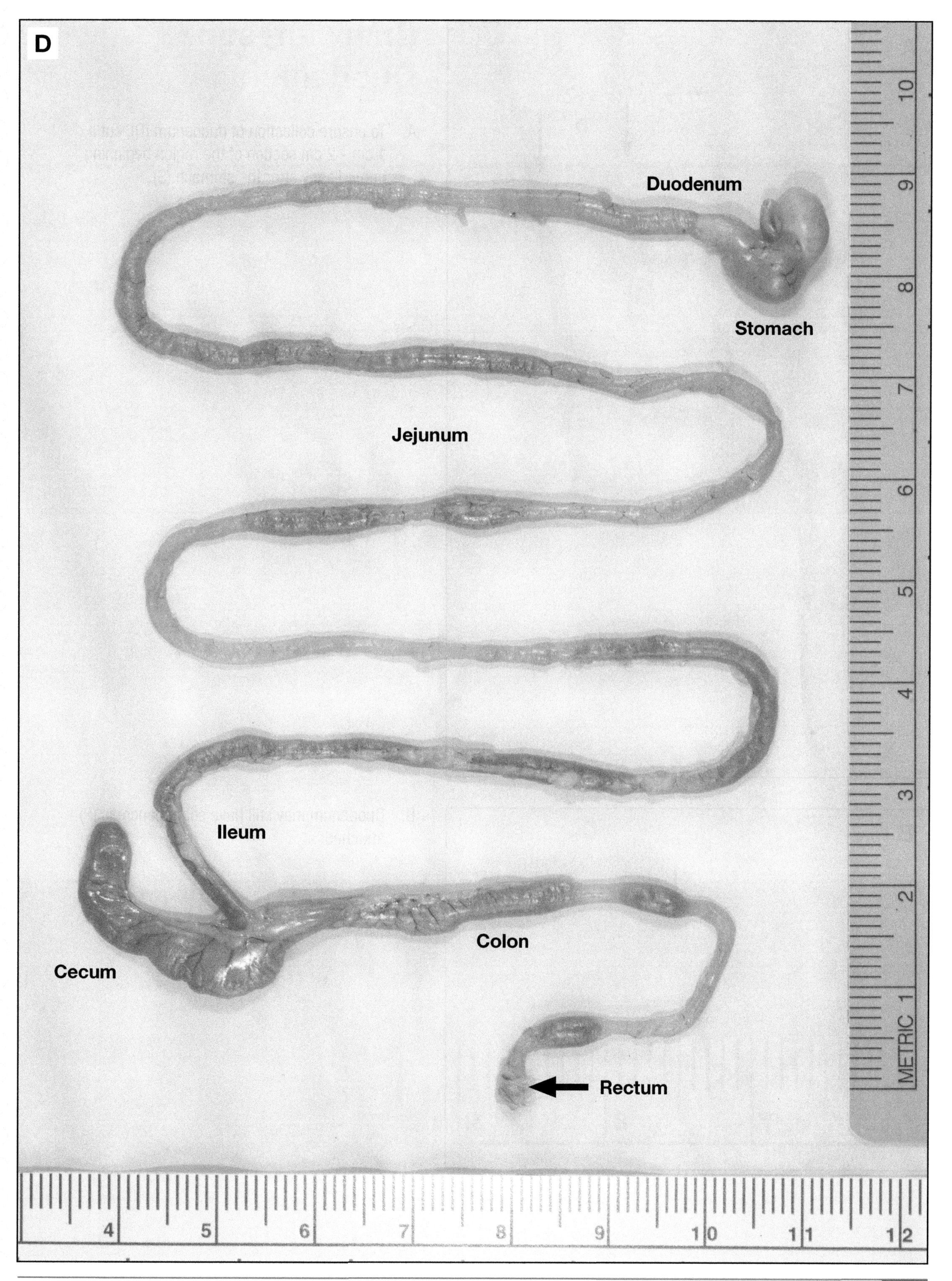
D
Duodenum
Stomach
Jejunum
Ileum
Cecum
Colon
Rectum
METRIC

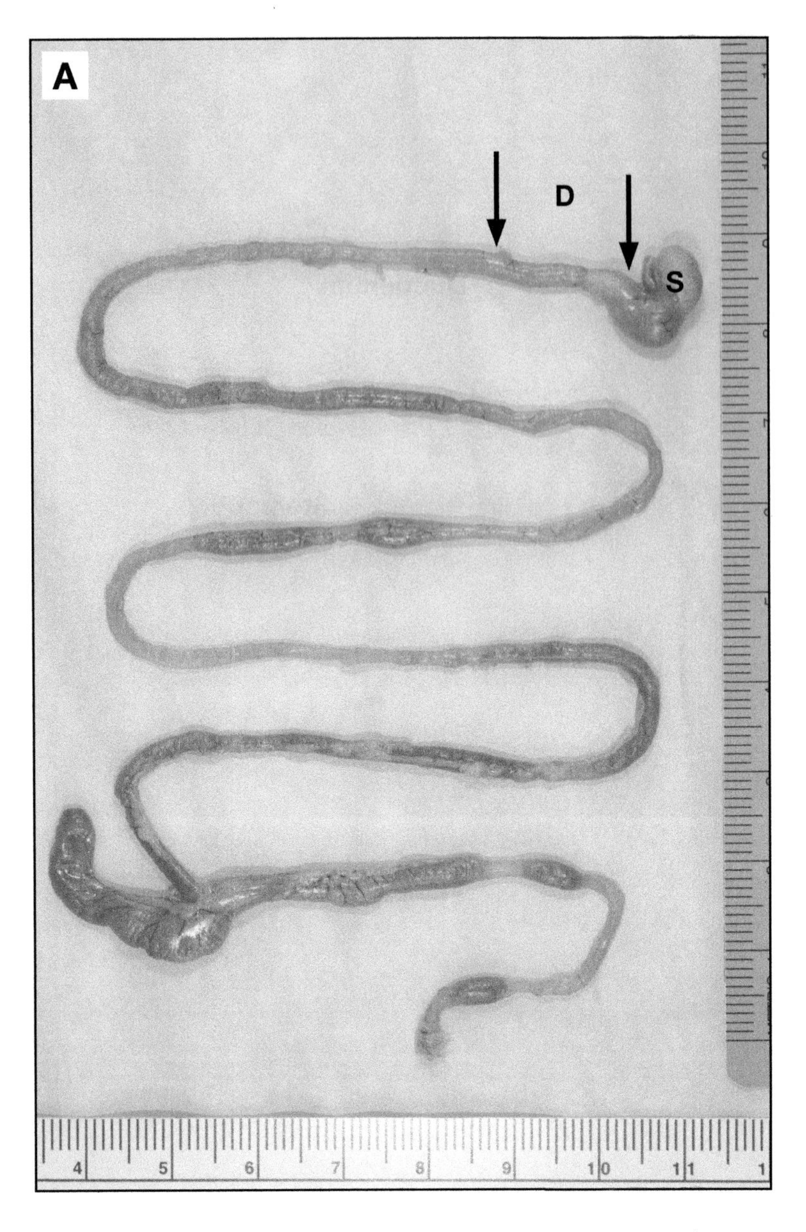

Small Intestine - Duodenum

A. To ensure collection of duodenum (D), cut a 1 cm - 2 cm section of the region beginning immediately after the stomach (S).

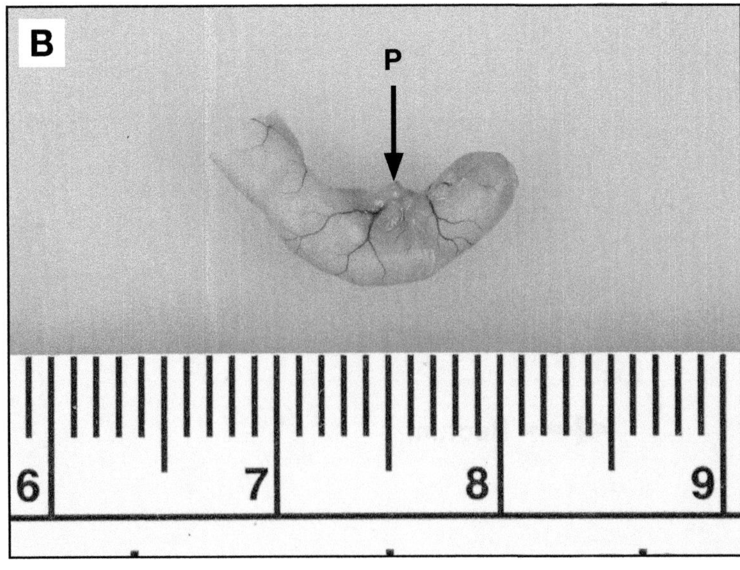

B. Duodenum may still have some pancreas (P) attached.

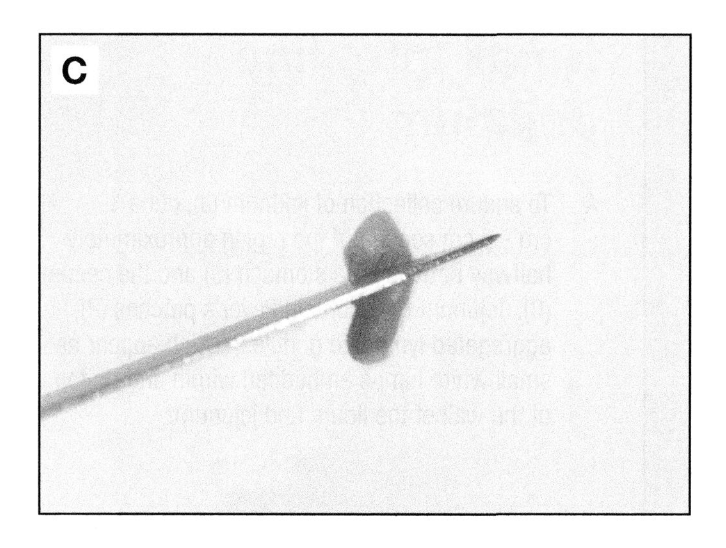

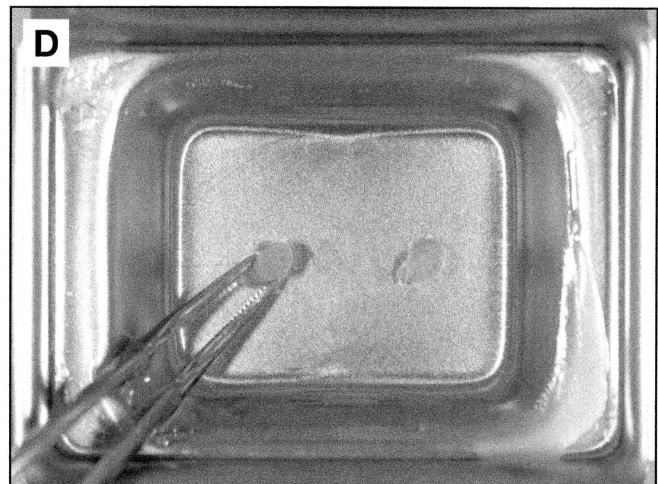

C. At embedding, cut the segment into 2 pieces.

D. It may be necessary to hold the duodenum segments vertically until the paraffin begins to harden. Embed to create a block with two cross sections of the lumens.

E. Face the block until a full cross section of each lumen is obtained.

F. The ideal section contains at least one complete, circular cross section of duodenum.

Single cross section enlarged to show detail.

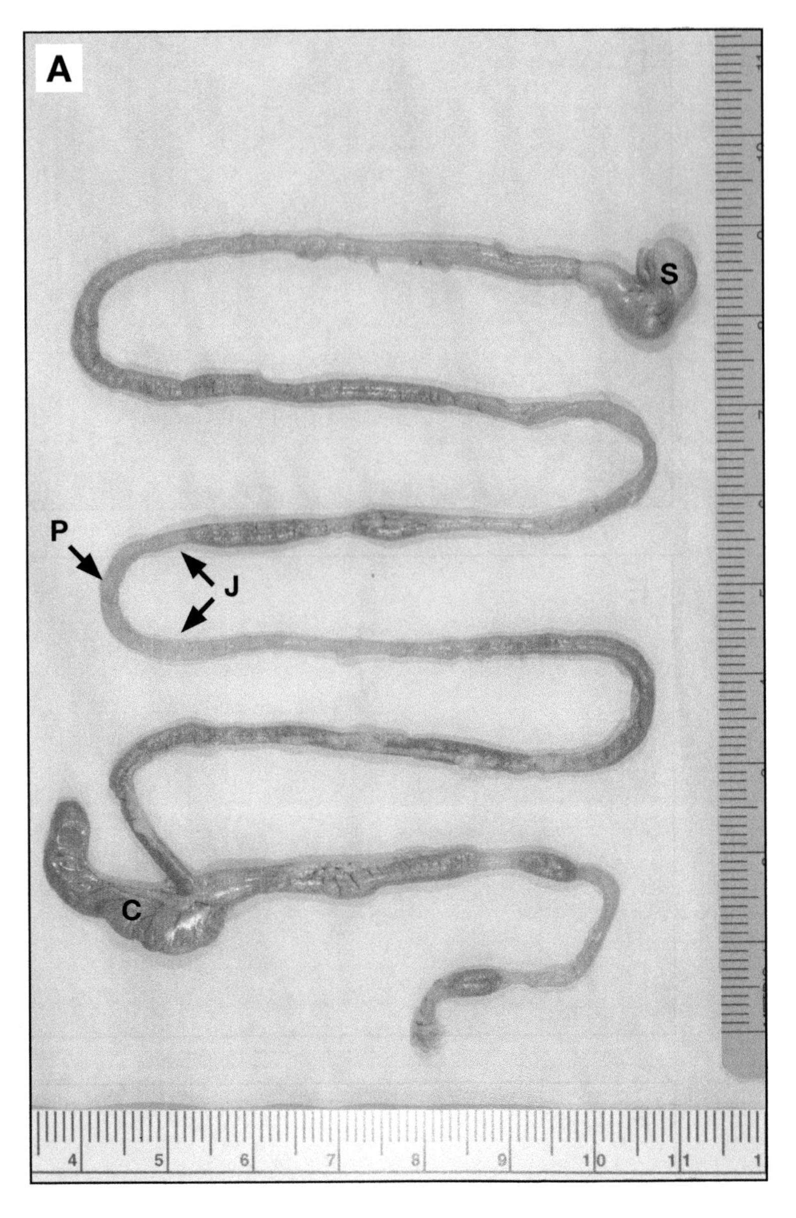

Small Intestine - Jejunum

A. To ensure collection of jejunum (J), cut a 1 cm - 2 cm section of the region approximately halfway between the stomach (S) and the cecum (C). Jejunum may contain Peyer's patches (P), aggregated lymphoid nodules, which appear as small white lumps embedded within and on top of the wall of the ileum and jejunum.

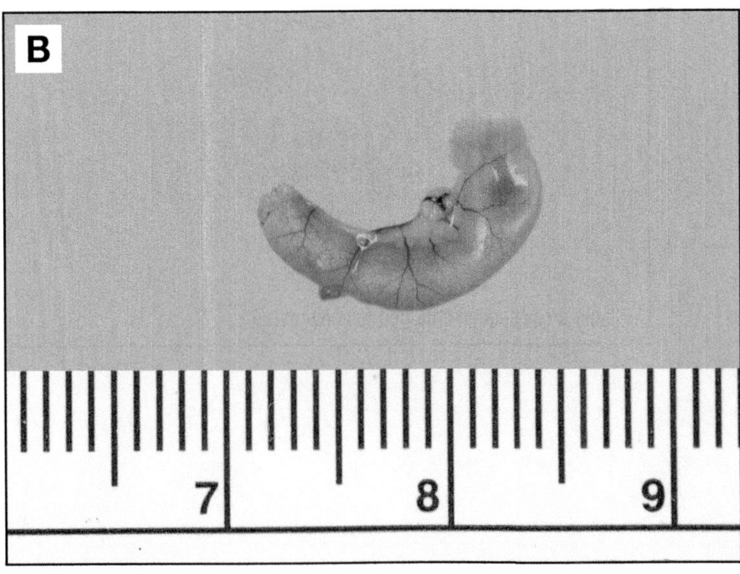

B. Jejunum and other intestinal tissues may curl slightly during fixation.

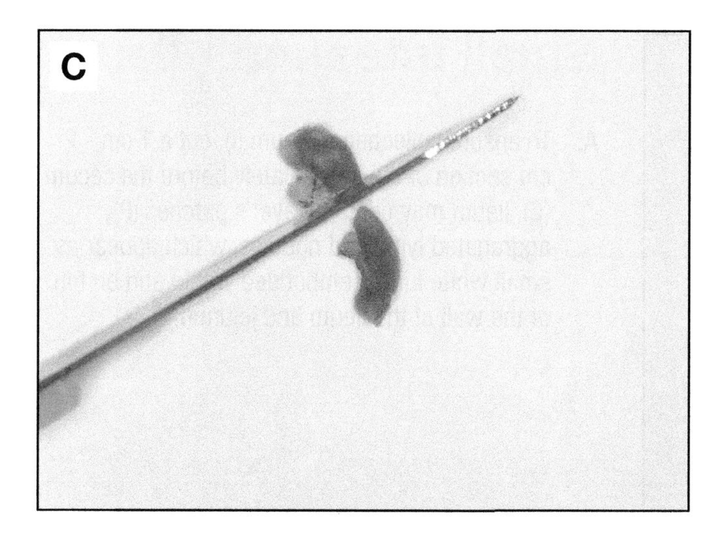

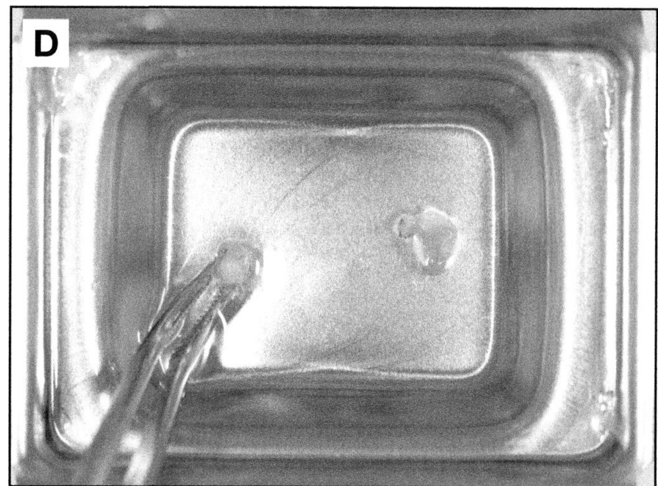

C. At embedding, cut the segment into 2 pieces.

D. It may be necessary to hold the jejunum segments vertically until the paraffin begins to harden. Embed to create a block with two cross sections of the lumens.

E. Face the block until a full cross section of each lumen is obtained.

F. The ideal section contains at least one complete, circular cross section of jejunum.

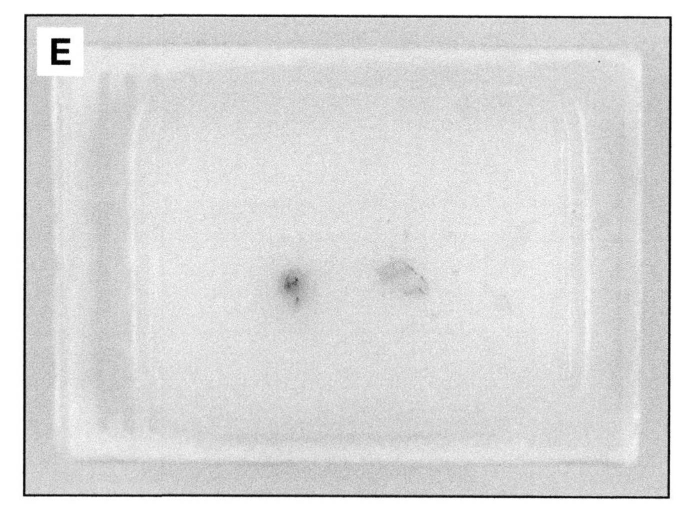

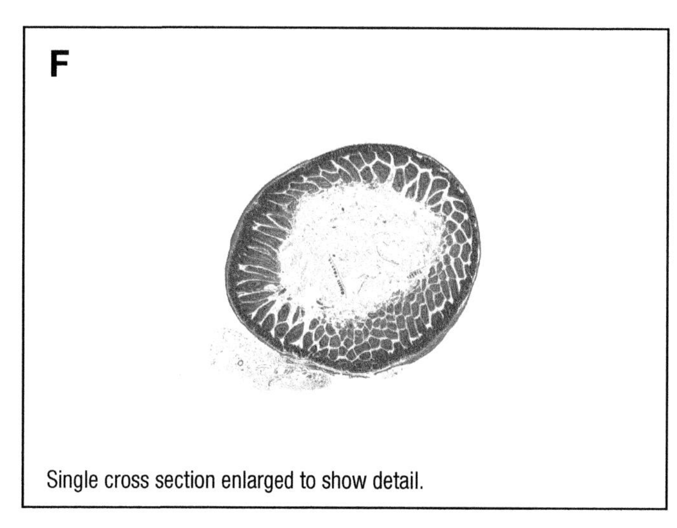

Single cross section enlarged to show detail.

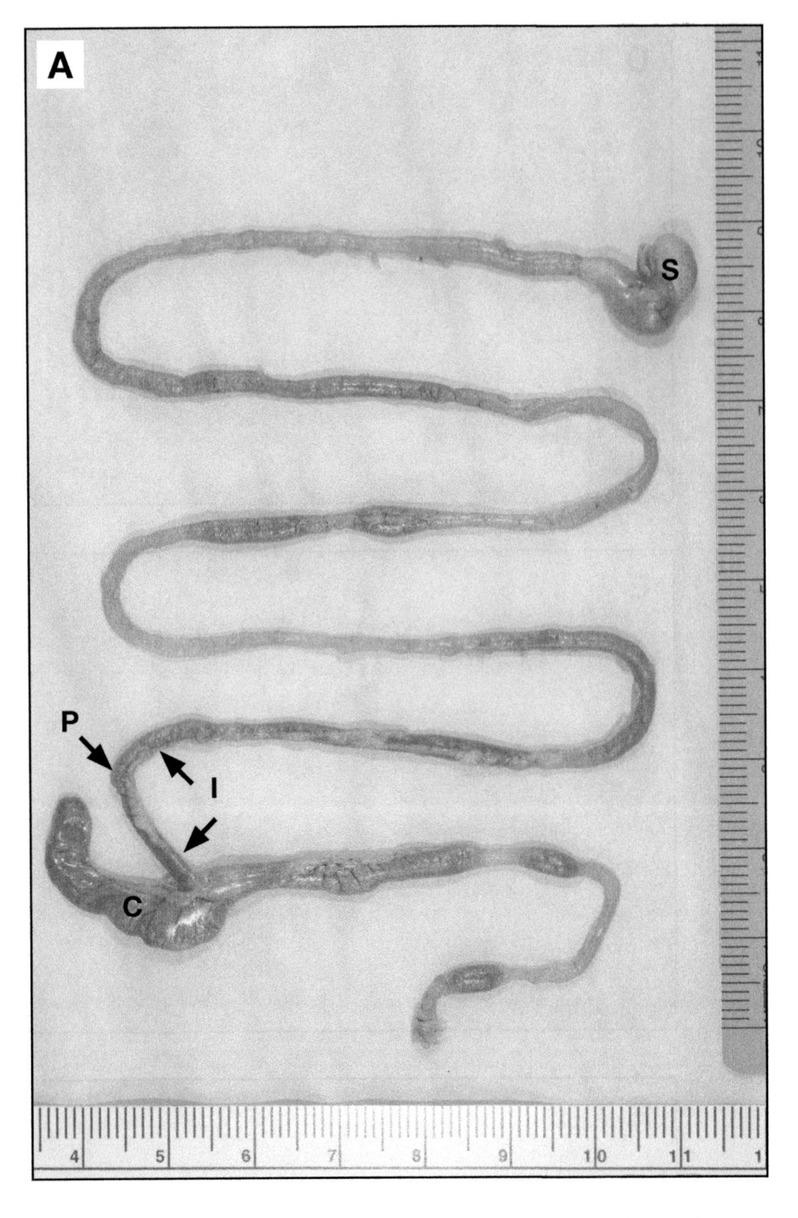

Small Intestine - Ileum

A. To ensure collection of ileum (I), cut a 1 cm - 2 cm section of the immediately before the cecum (C). Ileum may contain Peyer's patches (P), aggregated lymphoid nodules, which appear as small white lumps embedded within and on top of the wall of the ileum and jejunum.

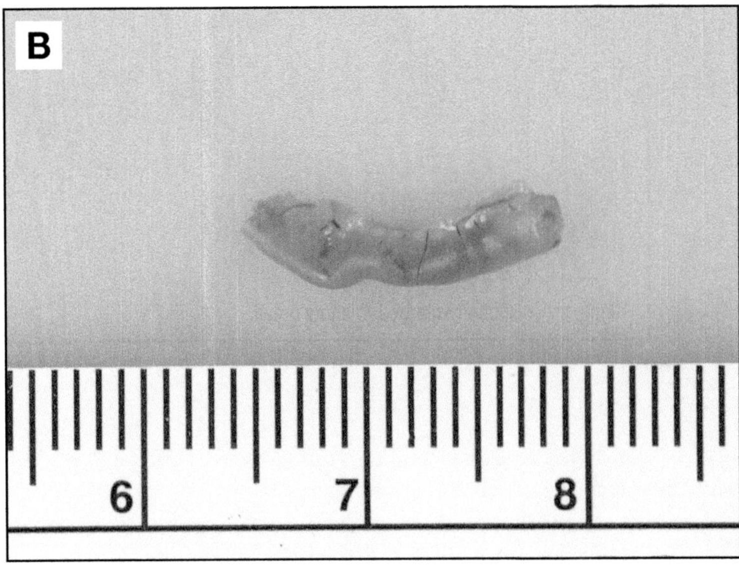

B. Ileum and other intestinal tissues may curl slightly during fixation.

C. At embedding, cut the segment into 2 pieces.

D. It may be necessary to hold the ileum segments vertically until the paraffin begins to harden. Embed to create a block with two cross sections of the lumens.

E. Face the block until a full cross section of each lumen is obtained.

F. The ideal section contains at least one complete, circular cross section of ileum.

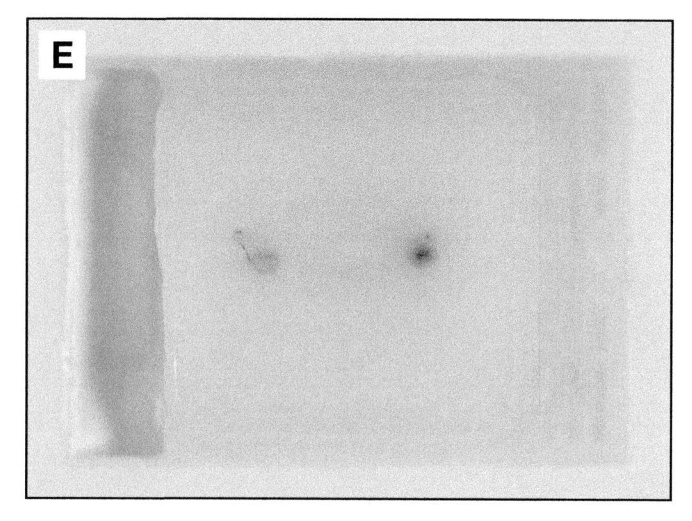

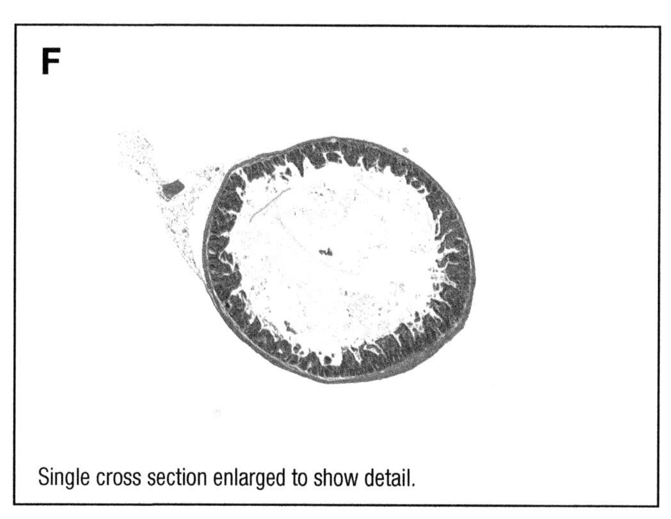

Single cross section enlarged to show detail.

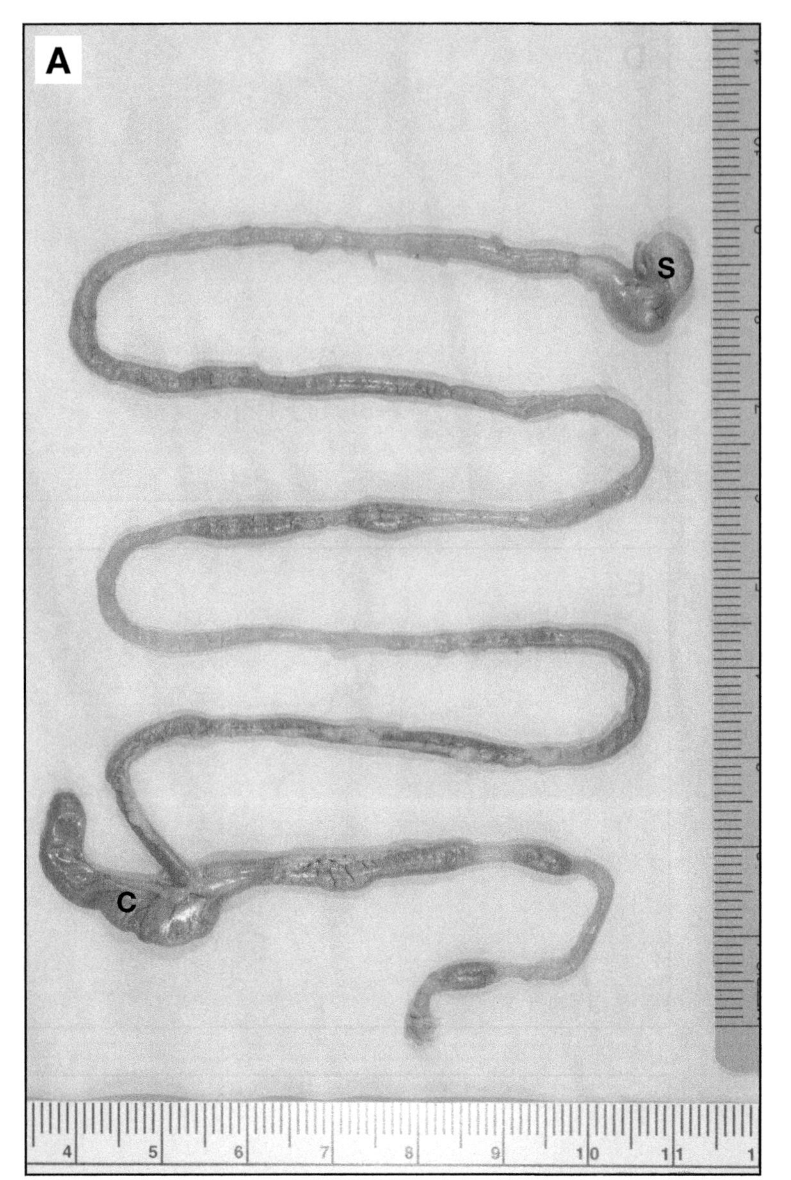

Large Intestine - Cecum

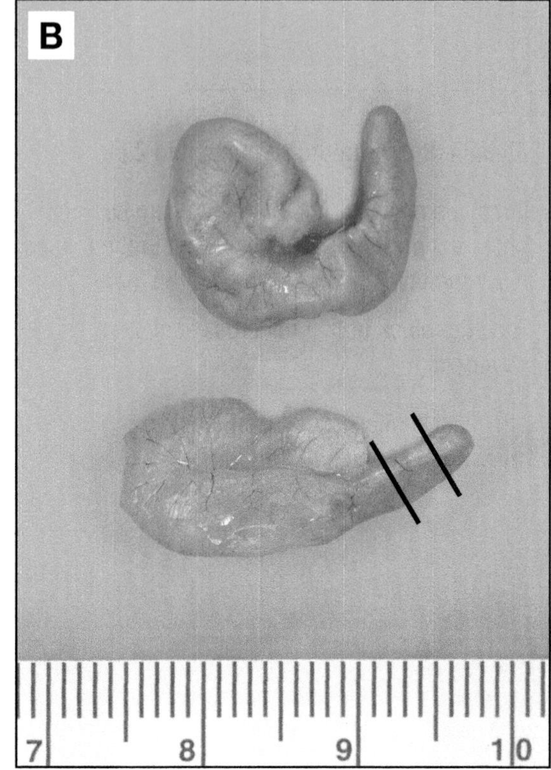

A. Collect the entire pouch of the cecum (C). The cecum may be rinsed out in a dish of saline to remove the contents.

B. Cecum is shown here unrinsed (top) and rinsed (bottom) of fecal contents. A cross section is taken across the lumen for evaluation.

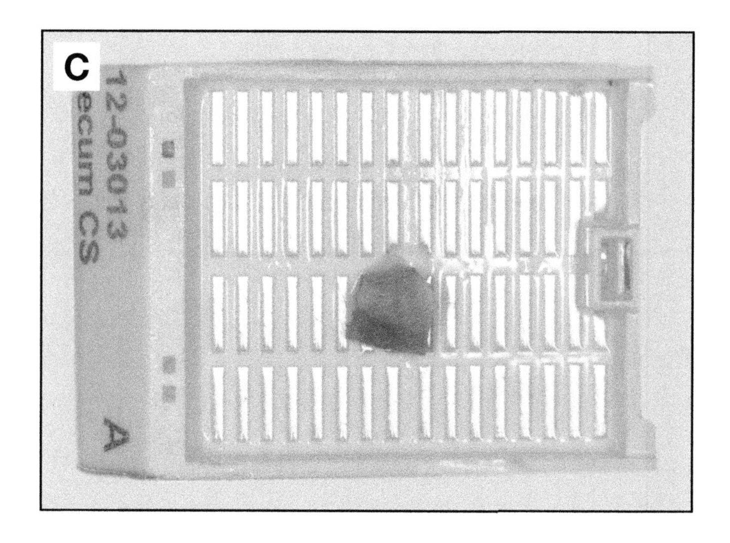

C. After processing, the cecum is smaller and darker in color.

D. Embed the cecum to obtain a cross section of the lumen.

E. Face the block until a full cross section of the lumen is obtained.

F. The ideal section contains a complete, circular cross section of cecum.

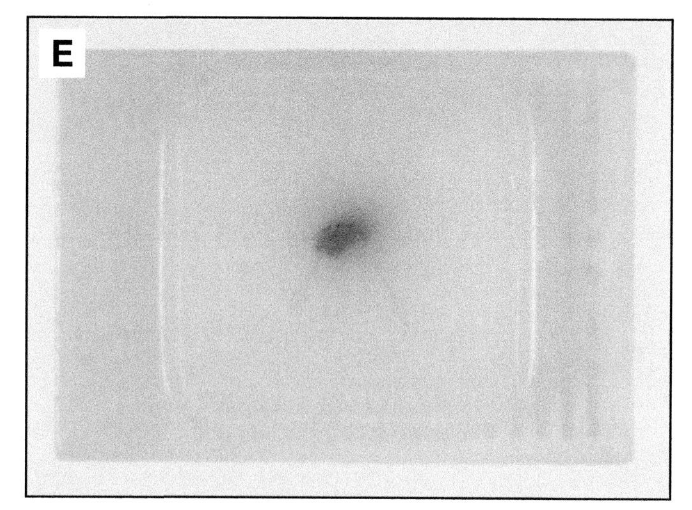

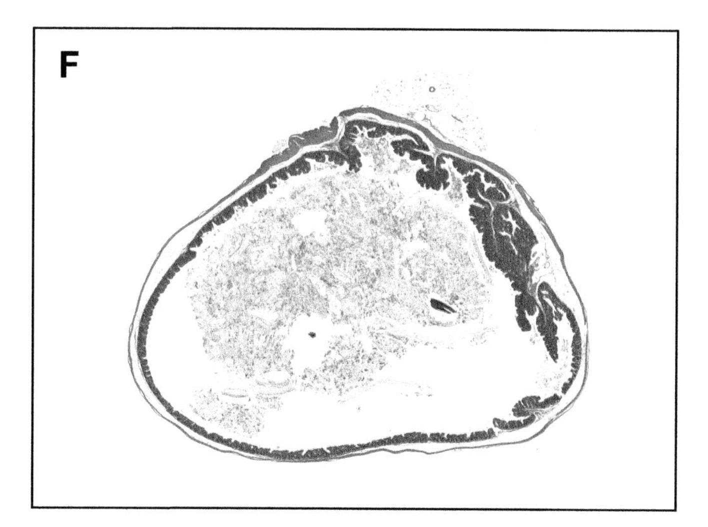

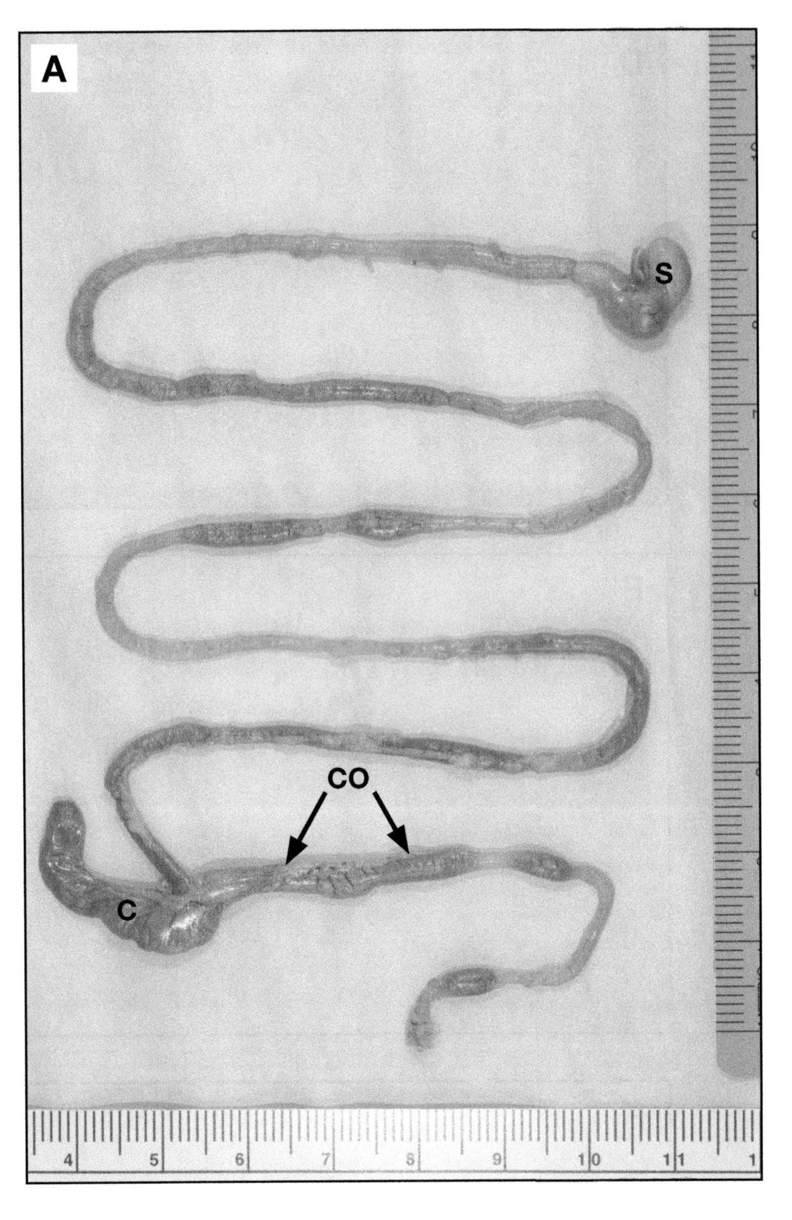

Large Intestine - Colon

A. To ensure collection of colon (CO), cut a 1 cm - 2 cm section of the region immediately after the cecum (C).

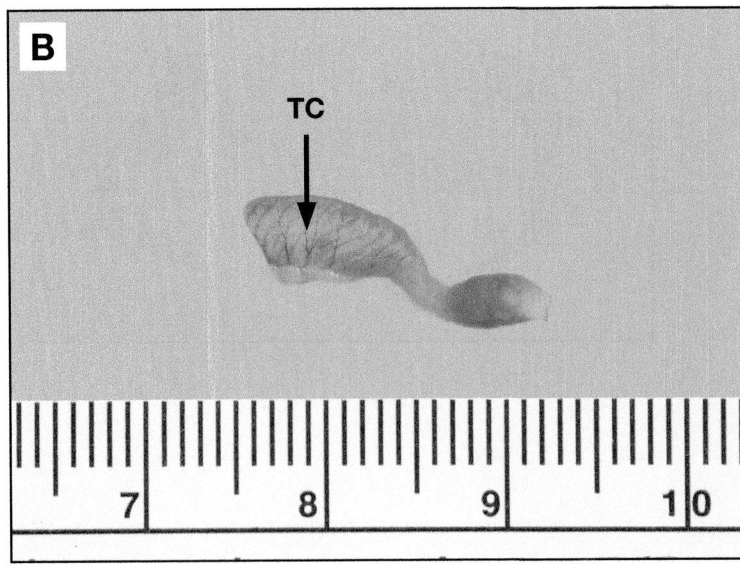

B. The colon has thin red lines i.e., tenia coli (TC), which wrap around this tubular area. Look for these lines to distinguish colon from other regions of the intestines.

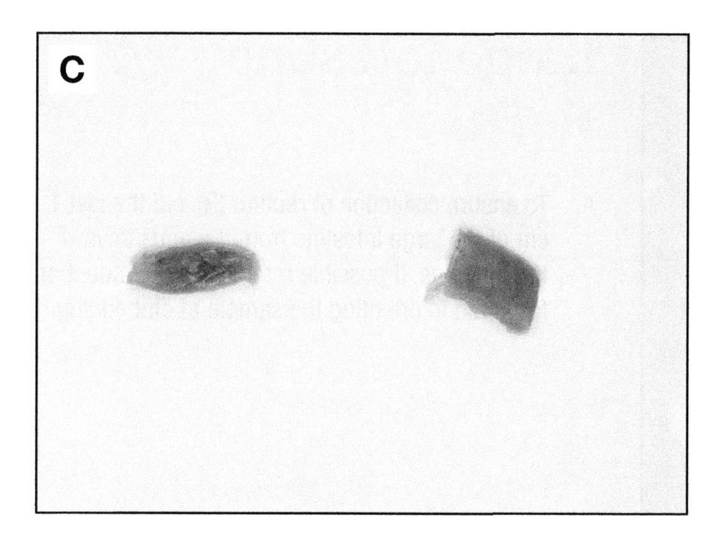

C. At embedding, cut the segment into 2 pieces.

D. It may be necessary to hold the colon segments vertically until the paraffin begins to harden. Embed to create a block with two cross sections of the lumens.

E. Face the block until a full cross section of each lumen is obtained.

F. The ideal section contains at least one complete, circular cross section of colon.

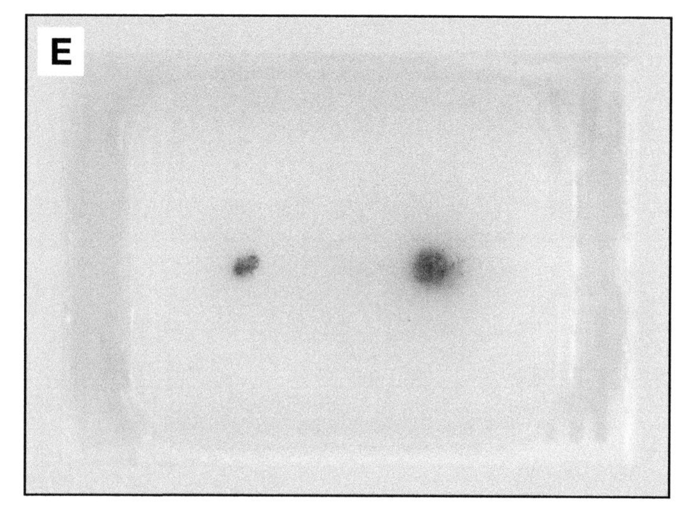

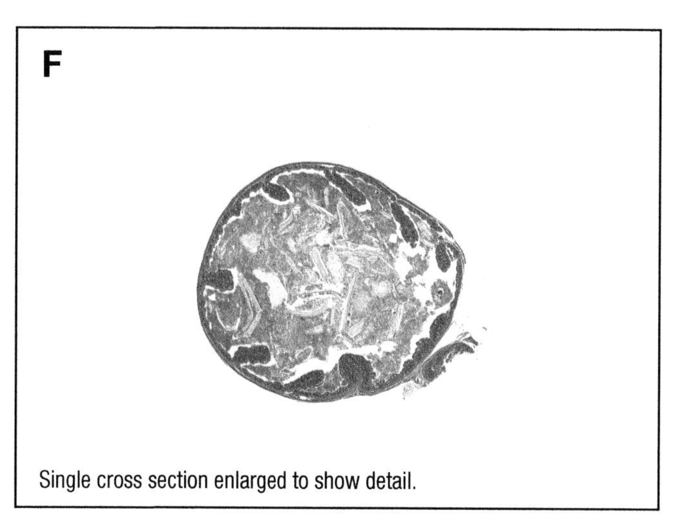

Single cross section enlarged to show detail.

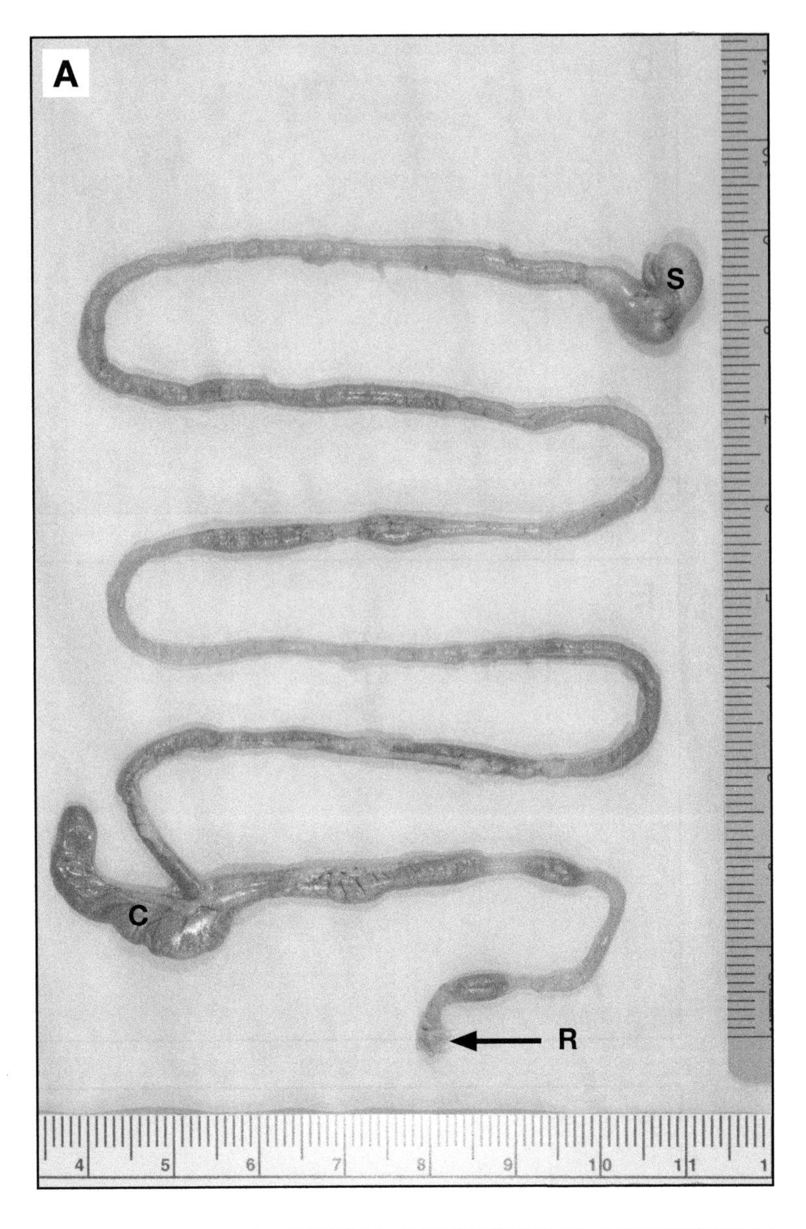

Large Intestine - Rectum

A. To ensure collection of rectum (R), cut the last 1 cm of the large intestine from the anus toward the intestine. If possible retain a small amount of fur to aid in orienting the sample at embedding.

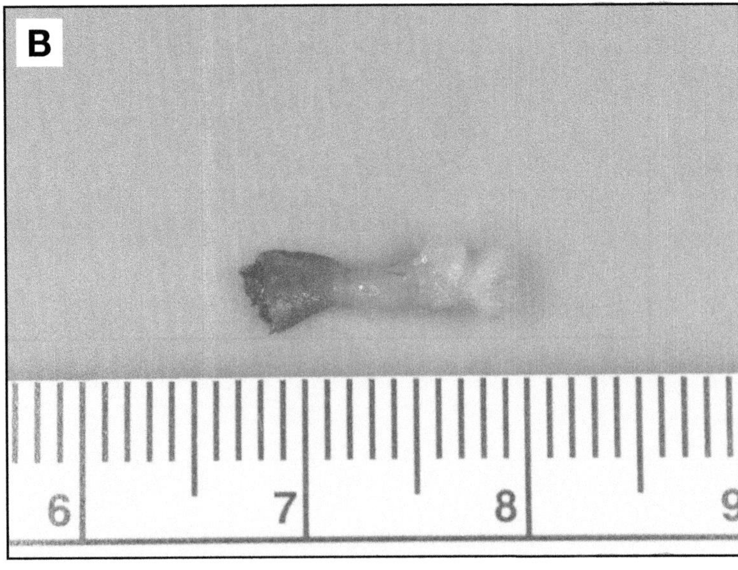

B. The rectum begins immediately after the distal colon. It is impossible to distinguish between the two regions with the naked eye, therefore the two regions are collected as a unit.

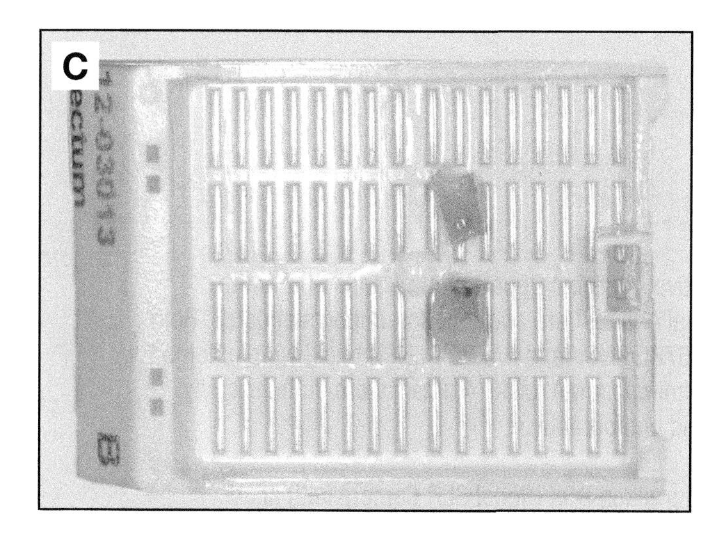

C. At embedding, cut the segment into 2 pieces.

D. It may be necessary to hold the rectum segments vertically until the paraffin begins to harden. Embed to create a block with two cross sections of the lumens.

E. Face the block until a full cross section of each lumen is obtained.

F. The ideal section contains at least one complete, circular cross section of rectum.

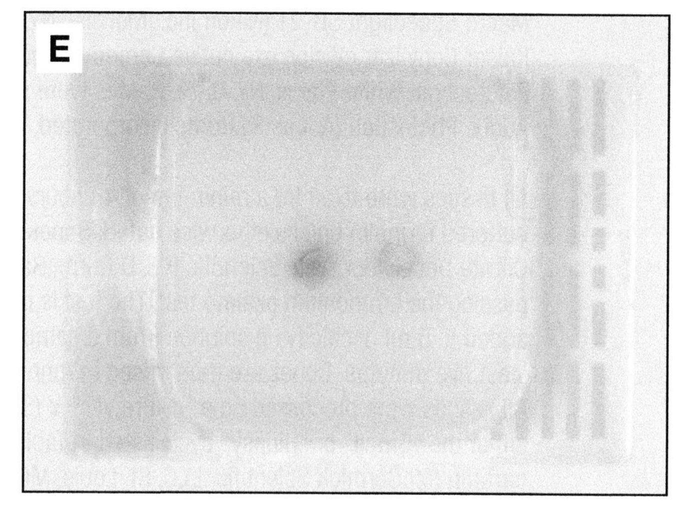

Single cross section enlarged to show detail.

Materials and Methods

All work with animals was conducted in the Comparative Medicine Department at Sanofi Genzyme (Framingham, MA, USA) in accordance with Institutional Animal Care and Use Committee standards. Both males and females of the two strains of mice used, C57BL/6NCrl and BALB/cAnNCrl mice, were obtained from Charles River Laboratories International, Inc. (Wilmington, MA, USA). Animals were euthanized by carbon dioxide asphyxiation prior to tissue removal and photography.

Necropsy and gross tissues were photographed using a Nikon D3X Digital SLR Camera with an AF Micro-Nikkor 60 mm f/2.8D or AF Micro-Nikkor 105 mm f/2.8D lens. Attached to either lens was a Nikon Macro Speedlight SB-21 (Nikon Inc., Melville, NY, USA). Additional lighting and copy work was done on a Kaiser Copylizer eVision exe.cutive Camera Stand with a Kaiser RB4.36 HF Copylizer Lighting Unit (Kaiser Fototechnik White Plains, NY, USA). Photos were shot in RAW format and post-production was done in Adobe Photoshop (Adobe Systems Incorporated, San Jose, CA, USA).

All tissues were fixed for a minimum of 48 hours, although most were fixed longer in 10% neutral buffered formalin unless otherwise noted. Bones were decalcified in Buffered Formic Acid Trisodium Citrate Decalcifier (Poly Scientific R & D Corp., Bay Shore, NY, USA) that was changed every other day until passing the ammonium oxalate test. The test is passed when 1 ml of 5% ammonium oxalate solution added to 5 ml decalcifying solution (from a sample being tested) remains clear of white precipitate after at least five minutes. Bones are then rinsed in running water for a minimum of one hour prior to processing. All tissues were processed on a Sakura VIP® V (Sakura Finetek, Torrance, CA, USA) Tissue Processor using either the "Small" or "Biopsy" cycle listed in Table 1 (page 90) and infiltrated with Paraplast X-TRA® brand paraffin (McCormick Scientific, LLC, St. Louis, MO, USA). Slow mixing was set on every step. Vacuum was only applied to the paraffin steps.

All sections were cut on a Leica 2055 microtome (Leica Microsystems Inc., Buffalo Grove, IL, USA) at a thickness of 5 micrometers. After drying all sections were stained on a Leica XL Autostainer with hematoxylin and eosin protocol as described in Table 2 (page 91). The slides were mounted with SHUR/Mount™ (Triangle Biomedical Systems, Inc., Durham, NC, USA) and coverslipped.

All samples were microscopically examined and imaged using a Zeiss Axio Scan.Z1 whole slide imager. Examination was conducted using a 20x, 0.8NA Plan-Apochromat objective. After image acquisition, the images were processed and exported using Zeiss Zen microscope software (Carl Zeiss Microscopy, LLC, Thornwood, NY, USA).

Table 1
Tissue Processing Cycles

Step	Solution	Small Cycle (Time in minutes)	Biopsy Cycle (Time in minutes)
1	40% Reagent alcohol	30	10
2	70% Reagent alcohol	30	10
3	70% Reagent alcohol	45	15
4	90% Reagent alcohol	45	15
5	90% Reagent alcohol	45	15
6	100% Reagent alcohol	45	15
7	100% Reagent alcohol	45	15
8	Xylene	45	10
9	Xylene	45	10
10	Xylene	45	10
11	Paraffin	20	5
12	Paraffin	35	10
13	Paraffin	35	15
14	Paraffin	45	15

Steps 1 - 10 were performed at room temperature with slow mixing. Heat (57°C) and vacuum were only applied to steps 11-14.

Table 2
Hematoxylin and Eosin Procedure

Step	Reagent	Time in Minutes
1	Xylene	3
2	Xylene	3
3	Xylene	3
4	100% alcohol	3
5	100% alcohol	3
6	95% alcohol	3
7	95% alcohol	3
8	Tap water	3
9	Hematoxylin (Richard Allan Hematoxylin I)	7
10	Tap water	1
11	Clarifier 2 (Richard Allan)	1
12	Tap water	1
13	Bluing solution (Richard Allan)	1
14	Tap water	1
15	95% alcohol	1
16	Eosin Y with Phloxine (Richard Allan)	1
17	95% alcohol	1
18	95% alcohol	1
19	100% alcohol	1
20	100% alcohol	1
21	Xylene	3
22	Xylene	3
23	Xylene	3

Richard-Allan Scientific™ Signature Series Hematoxylin 1, Clarifier 2, Bluing solution and Eosin Y with Phloxine. (Thermo Fisher Scientific Inc., Kalamazoo MI, USA).

References

- Treuting P.M. and S.M. Dintzis. 2002. Comparative anatomy and histology: a mouse and human atlas. Oxford: Elsevier Publishing.
- Carson F.L. and C. Hladik. 2009. Histotechnology: a self instructional text. 3rd edn. Hong Kong: Clinical Pathology Press.
- RENI tissue trimming guide (Internet). Available from : http://reni.item.fraunhofer.de/reni/trimming/index.php
- Brayton C. 2001. Evaluating mutant mice: anatomic pathology. Vet Pathol 38:1-19.

9780367177751